ECOLOGY AND EARTH HISTORY

BIOLOGY AND ENVIRONMENT
General Editor:
Richard N. T-W-Fiennes

THE LIVING OCEAN: MARINE MICROBIOLOGY
E.J. Ferguson-Wood

TERRESTRIAL ENVIRONMENTS
J.L. Cloudsley-Thompson

MAN AND WILDLIFE
L. Harrison-Matthews

WATER AS A PRODUCTIVE ENVIRONMENT
C.F. Hickling

MAN AND NATURAL RESOURCES: AN AGRICULTURAL PERSPECTIVE
Sir Cedric Stanton Hicks

Ecology and Earth History

RICHARD N. T-W-FIENNES

CROOM HELM LONDON

First published 1976

Croom Helm Ltd.
2-10 St. John's Road, London SW11

ISBN 0-85664-038-7

Printed in Great Britain by Biddles Ltd., Guildford
Bound by Redwood Burn Ltd., Trowbridge and Esher

CONTENTS

INTRODUCTION

This volume outlines in simple terms the basic concepts of ecology. It is not a textbook of ecology, but an introduction to the subject and to the more advanced presentations of special aspects that will appear in succeeding volumes. There are many excellent textbooks of ecology, to some of which the reader is referred and to which the present volume should also serve as a useful primer.

I have linked this presentation of ecological theory to a survey of present theories of Earth History. All of earth history is, in a sense, ecology. Earth's environments have evolved throughout the ages from the time when no life existed, just as life forms have evolved from primitive to complex. It seems to me, therefore, that one should understand how these environments have come into being. I have also presented a simple account of the energy systems that are used by living things. These too need to be understood, if we are to appreciate energy problems. Man, like all other living things, requires the energy provided by his food, but unlike other creatures his way of life demands additional ('auxiliary') energy, and this has introduced a new factor to man's ecology. Hitherto, most of this auxiliary energy has been obtained from fossil fuels, coal, oil and uranium. But these are not inexhaustible; they represent capital, not income, and it is still not clear what the sources of the auxiliary energy will be.

The sun's energy is responsible also for the movements of the clouds, the winds and the ocean currents, and for circulating fresh water from the oceans to land surfaces. Otherwise, there could be no life on land. Here again, man's ecology demands more than that of other animals. Not only must man have water for drinking, but also for removal of his industrial and urban wastes. There are signs too that, lacking further energy input, this resource may be inadequate.

Man's future is beset with many doubts and complexities. To overcome them, mankind must understand his nature and move forward to find solutions that will be of benefit to all.

NOTES

1 J.C. Wylie, *Fertility from Town Wastes,* Faber, London, 1955

2 H. Godwin and A.G. Tansley, *The Natural History of Wicken Fen,* Bowes & Bowes, Cambridge, 1929

3 R.N.T -W - Fiennes, *The Ecology of the Grasses of the Lango District of Uganda, East Africa,* Government Printer, Entebbe, 1939

4 H.W. Mulligan (ed.), *The African Trypanosomiases,* Allen & Unwin, London, 1970
5 R.N.T -W - Fiennes, *Man, Nature and Disease,* Weidenfeld & Nicolson, London, 1964

Further Reading

Bertram, Colin, *Adam's Brood,* Peter Davies, London, 1959
Cloudsley-Thompson, J., *Animal Conflict and Adaptations,* Foulis, London, 1965
Reed, Laurence, *An Ocean of Waste,* Conservative Political Centre, London, 1972

1 GENERAL PRINCIPLES OF ECOLOGY

The Evolution of Habitat

In its simplest meaning, ecology is the study of habitat (Greek *oikos,* house or habitation). Ecology has, however, become a complex science with a terminology of its own and journals devoted to it. Only recently has the word appeared in the popular press in connection with environmental problems. In spite of this, as a science it has a respectable ancestry. Early naturalists, including Wallace and Darwin, recognised the relationships of living things with their habitat and with each other. Definitive studies on ecology appear to have been initiated by a German scientist, Möbius, in the 1880s. Subsequently and until the present day, the importance of such studies has been widely recognised and they have been pursued in Britain and North America, as well as in Germany. A critical review of the subject is given by Macfadyen,[1] and a short, but useful account by Healey.[2]

It is obvious that certain kinds of animals or plants can only live in certain types of habitat. Fishes cannot live out of water and a monkey thrown into water will drown if it cannot escape. It is generally accepted, too, that since life appeared on earth there has been a progression of life forms of increasing complexity. This progression is attributed to evolution, a process which increasingly adapts living things to be successful in the areas where they are evolved.

It may be less obvious that not only life forms undergo a process of evolution, but habitats as well. You have only to observe the local duck-pond to see that, left unhampered, water weeds grow in it and in time it gets filled and becomes earth with grass and trees and shrubs growing where there was once water. Similarly, look at any disused road or air strip and you will see that the surface is gradually weathered away, plants appear, in time soil is formed and it will eventually appear not unlike the surrounding area. Even on hard concrete, you can see that lichens quickly appear on the surface, followed by mosses and other plants.

Ecology basically looks at habitats in an evolutionary sense. It regards them initially as sterile areas, devoid of vegetation or other forms of life. 'Pioneer' forms of plant life then appear and begin to erode the surface or alter it in some other way which makes it more suitable for colonisation. The process continues and more demanding plants appear, which continue the process. At the same time, weathering assists the development: wind and rainfall break up and fissure the surface; frozen water in winter widens the fissures; roots of shrubs or small trees will enter the fissures and break them wide open; and soil begins to be formed. The evolution of a habitat proceeds by a series of steps or 'successions' until the 'climax' is reached. The 'climax' is the point at which both habitat and life forms on it are in full unison, so that development will not proceed further provided there is no change of climate. Examples of 'climax' habitats are primary forest, such as the big equatorial forests in Central Africa and South America, and prairie grasslands like those developed on chernozem soils across central North America and the steppes of Siberia and southern Russia into Europe. In such areas, a characteristic animal fauna is developed also, adapted to the conditions of the habitat: there are forest animals and birds in the forests, and plains animals and birds on the plains, differing in their ways of acquiring food, in colour and other protective measures, and in their general reactions and physiology.

The general process of 'succession' is known as a 'sere'. The process arising in water is known as a 'hydrosere' and that on land or rock surfaces as a 'xerosere'. The direct succession to the climax is known as the 'primary succession'. If, however, there is interference with the climax, as by cutting timber or clearing for agriculture, a 'secondary succession' develops, which may be a reversion to an earlier stage of the 'primary succession', or something different. When there is interference with a pre-climax stage of a succession, as, say, by dry season burning, the 'primary succession' may be deflected, and we speak of a 'deflected succession', which may reach a kind of climax of its own persisting until the cause of the deflection is removed.

An understanding of these seres and their stages is fundamental to an understanding of environmental problems and pollution by human activities. Most of these create a new form of pioneer surface, on which the object of reclamation is to recreate an advantageous succession leading to recovery in the shortest possible time. Soil erosion leads to loss of topsoil and in tropical areas to laterisation – that is chemical change and rock-like concretion – of the subsoil. The reverse process under natural conditions may take centuries, but the ecologist may find ways to hasten the process. Large sections of the Sahara carried human and animal populations within human memory and have been converted to desert largely by human activities.[3,4] If the process of desiccation can be reversed, this can only be done with a knowledge of the ways in which suitable ecological successions can be encouraged. A polluted river or estuary is another instance of a pioneer type of substrate, which can be remedied if the right kind of microbes are introduced to devour the pollutants or in the case of chemical pollutants to alter them by their metabolic activities. Even oil slicks can be attacked by the right kind of organisms and quickly degraded into harmless substances or even into nutrients for higher forms of life.

The spate of journalism in recent years on ecology and the environment appears to ignore the absolute necessity of studying the ways in which natural processes may be stimulated and accelerated. Governments appear mesmerised by the necessity of appearing to do something, even the wrong thing, rather than to invest the public's money in fundamental research on these problems. One might suppose, from what one reads, that human activities are universally detrimental to earth's environment, certainly from his own point of view. The natural successions have been either deliberately or unwittingly 'deflected'. The problem is to understand the nature of these deflections and to remove their undesirable features; this means study.

The Ecological Community

Hitherto, we have considered the successions which occur in an ecological sere. Let us now look at the concept of the 'community' or 'ecosystem', to use the term introduced by A. G. Tansley[5] in 1935. First, what are the factors which determine the type of community to be developed in a given habitat? There are only two of major importance, the climate and the soil. It might be objected that this ignores the importance of the atmosphere. The composition of the earth's atmosphere has, however, been so constant over millions of years that we can for the present ignore it.

Like much in ecology, the influence of climate is superficially obvious. In the most extreme climates, waterless desert and polar ice, life cannot exist, although abundant unicellular protozoa live in the waters below the polar ice. At the other extreme, life is most abundant and prolific in the wet tropics, where rainfall is heaviest and the sun's rays are most powerful. In such areas, too, seasonal variations are at a minimum, whereas in other areas rainfall and temperatures are variable throughout the year and the vegetation varies also. Both plants and animals show adaptations to lean periods when water is scarce or the temperatures low. Trees shed their leaves; annual plants bury their seeds in the ground to grow again when conditions are favourable; some birds – mammals too – migrate, and some mammals, birds and reptiles hibernate or aestivate. These are commonplace facts and do not require elaboration. They are, however, facts of which ecology takes account, and which the ecologist must study if he is to understand his area.

The facts relating to soil conditions are somewhat more complex. As a basic principle, soil is formed from a parent rock, usually the rock underlying it. However, it may be impoverished by leaching or erosion, or enriched by wash or silt. Usually the chemicals present in a soil will only be those present in the parent rock. If this is deficient in any of the minerals important for plant growth, these will be deficient in the soil too. There are many minerals important for plant growth, such as calcium, phosphorus, magnesium, zinc, cobalt, molybdenum, manganese, iron and copper. In some cases, only trace quantities of the element are needed. In some areas, plant growth is limited by these deficiencies and is less luxuriant than climatic conditions would appear to warrant. As a result of the poor growth of herbage, the development of a rich soil is either retarded or inhibited. It also happens that soil minerals may be adequate for the development of a rich plant cover, but are inadequate for the nutritional requirements of animals that feed on them. In this case, the fauna is scanty, although the vegetation is rich and might be expected to support large numbers of animals. In all parts of the world, areas are known where domestic animals do not thrive on the pastures, suffering from deficiency diseases known by various names, such as pine, swayback, staggers, and so on. These diseases were for a long time a mystery to stock-breeders, until it was discovered that there existed underlying defects in the minerals present in the soil. The deficiencies are sometimes slight, and one Australian farmer remedied a deficiency of copper by drawing a short length of copper pipe across his pastures behind a tractor.

As soils develop, they become enriched by the addition of humus. This is derived from broken-down organic debris of dead plant and animal material. Thus, the more quickly rich vegetation becomes established in a habitat, the more quickly will the soil become improved and move towards climax conditions. Here again the ecologist, and indeed the town-house gardener, can accelerate the process by thorough cultivation and by addition of materials to the soil to remedy deficiencies and improve its structure and water-holding capacity. These activities, however, must be properly orientated and based on proper soil analysis and examination. With the more difficult types of habitat, such as sand dunes and salt marshes, much can be done also by introducing pioneer plants adapted to them; certain types of grass will grow on sand dunes and, when introduced, will initiate the soil formation process.

Quite apart from the plants we can see, the soil itself teems with life. Everybody knows of the beneficial effect of earthworms in aerating and improving the soil structure. The soil is also alive with bacteria, fungi, small helminths and other organisms, whose activities are advantageous; indeed, without this active life in the soil itself, many of the higher plants we can see could not exist. Not least amongst these are the small fungi and bacteria which form root nodules on coniferous trees and leguminous plants, such as clover, so fixing atmospheric nitrogen, and their allies in the soil.

Also amongst these organisms are the so-called saprophytes,

responsible for breaking down dead organic material and making its products ready for re-use. Without this 'recycling' of organic materials, life will in time grind to a halt. Unfortunately, in the euphoria which followed the discovery of agricultural chemicals, their effect on soil communities was largely overlooked as with their effect on the insect life required for plant pollination. Today, in our large combine-harvested fields scant attention is still paid to the effects of these chemicals on the soil-living communities; pests are eradicated by chemicals and deficiencies made good by the use of artificial fertilisers. Aeration and structure of the soil are ignored and experts are asking how long such conditions can persist; belatedly land-owners are stipulating specified rest periods for their fields during which recovery can take place. But can it? Again, organised ecological studies of soil life are essential, if permanent deterioration of soils is not to take place as has happened over such vast areas of the American Middle West and elsewhere.

Soils, therefore, to achieve optimum development must not only contain all the minerals necessary for the plants and animals they are to support, but must also be properly aerated and maintained by the small organisms that live in them. These organisms form the base of the ecological pyramid, their biomass (weight of living matter) far exceeding that of the visible plants and animals which depend on them at higher levels. Many of them use energy systems quite different from higher plants and animals; some do not require oxygen; some use as energy sources chemicals that are not available to higher plants and animals. Thus they perform activities that cannot be undertaken by other life forms and in situations not available to them; for instance they are found both in oil wells and in the upper layers of the atmosphere. They provide us with our antibiotics, our wine, bread, beer and soft cheeses, and are used in many industrial processes. To understand their activities is as important as not destroying them by artificial processes in agriculture.

An ecological community, then, is an association ('associes') of micro-organisms and higher plants and animals, which live together, are dependent on each other, and contribute to the improvement of the environment to a point dependent on conditions of climate and soil. We can now study the organisation of the community and the factors which lead to its stability. This is of special importance because man's activities in one way or another have so radically altered so many habitats and communities in ignorance of their structure and indifference to the effects. This is no new problem, because man has been doing it for 10,000 years or more; it is a pressing problem because of the increase in human numbers which has made the effects more widespread.

The Stability of the Ecosystem

Most plants and animals, of which the rat, man and the housefly are notable exceptions, are very 'habitat specific'; that is to say that they are confined to habitats which suit them within rather narrow limits. Quite slight alterations of a habitat result in a different species of bird, for example. When the habitats merge or intergrade, there may be a 'tension zone' where the two species are found but are in competition. Such gradations are seen in the case of a 'hydrosere', where at one end there is open water gradually rising through marsh from sloping land to upland. In such a situation, the vegetation – and the animals dependent on it – is 'zoned' from aquatic plants through marsh plants, to lowland wet-loving plants – grass and sedge – to dry-loving upland plants and trees.

A habitat comprises numerous 'niches', and there is a life form ideally suited to fill each niche. If a niche is empty, some form will quickly appear to fill it. Niches are extremely varied. We have already seen how many varied organisms live in and on the soil. Small insects and other creatures live in cracks in the bark of trees, under stones, and in every conceivable place where they can achieve their needs for food and breeding and reasonable security from predators. The term (niche) covers not only the places in which these creatures live, but also food requirements. For example, a niche is offered by animals living in the bark of trees for predators such as woodpeckers, or monkeys with long thin fingers that can get at them. Thus in each habitat, there exist food chains or food webs, which can be exceedingly complex in rich habitats such as tropical forest or rather simple in poor habitats such as desert. Basically, the energy which is the driving force of life is trapped by plant chlorophyll and used to convert carbon dioxide (CO_2) into higher carbohydrates such as starch, sugars, and fats. The stored energy is passed from mouth to mouth, through herbivores, carnivores, predators and scavengers, until it is dissipated by processes of decay and the residual organic materials are returned to the pool to be recycled.

In fertile areas, the communities developed are complex, well balanced and stable. Obviously they cannot be stable unless the numbers of plants and animals living in the community are kept within the limits that the habitat can support. To achieve this, the natural balances are maintained at optimum levels by a number of devices which serve to limit numbers when they tend to become excessive, and to build up numbers when conditions become favourable. Either way, the result is achieved in a remarkably short time. The underlying principle of this control is simple. More young are produced than can survive and the excess are eliminated in one way or another. Since the strongest – or at least the most fitted to the conditions – survive, the tendency is for constant improvement of the stock or a change of characteristics if this is advantageous because of changed circumstances. Once this principle is understood, it is easy to see the basic difficulties underlying human population problems at the present time. It is our endeavour to have as many of our babies survive as possible, irrespective of population pressures, and we thus preserve the weak and ill-adapted as well as the strong and well adapted. In this way, we could on the face of it be contributing to deterioration of the race. Authorities, such as the British geneticist Dr A.R. Penrose,[6] think that this is not necessarily the case, and that evolution in human communities is powerfully at work, but in a different way, as discussed in Chapter 4. J.B.S. Haldane also held this view.

Elimination of excessive young, which we shall examine below, is not the only way in which numbers are controlled. Both mammalian and avian populations are very sensitive to population pressures; when populations tend to become excessive, the community enters a 'stressed' state, especially if food is short. The condition of 'stress' is undergone by all members of the community and depends on alterations in the activity of glands of the endocrine system, especially the cortex of the adrenal gland. Through complicated feedback mechanisms, the animals become psychologically affected and more aggressive, the breeding urge becomes diminished and the females when they become pregnant are inclined to abort, and in any case bear fewer young. The young, when born, are inclined to be attacked and killed, and older and sick animals are also killed. In addition, the population becomes more susceptible to endemic diseases and parasites so that there is a higher natural death rate. In some animals, a migratory instinct develops and large numbers may leave the habitat to seek other suitable areas, or perish in the attempt. The suicidal migrations of the lemmings in Scandinavian countries are the best known example.[7]

The extent to which animals produce excessive young varies greatly. It is probably greatest amongst parasites, such as ticks and helminth worms, which must find a new host by rather hit and miss methods. Thus a female tick produces millions of eggs, dying herself in the process; these all hatch into larvae, of which probably only a few score find hosts; incapable of a free-living life, the rest perish. Small, vulnerable, short-lived animals such as rats and mice tend to produce large numbers of young, of which only a few survive unless conditions are especially suitable, when a plague of them may occur. The smaller birds too produce many more eggs than are likely to be needed for the next generation of breeding stock.

A number of these excess young may die at the hands of predators or from failure to find sufficient food. There is, however, a further controlling mechanism of the greatest importance, not only to the control of populations, but also to the structure and behaviour of a population within the community. This is the 'territorial' instinct. The instinct for territory is extremely ancient and universal, at any rate throughout the vertebrate kingdom. It is present among fish as well as mammals. It is surely not defunct in man, though not universally operative. The underlying principle, as with most principles of ecology, is again a simple one. Animals do not pair and breed until they have acquired a territory, and the territory they acquire must be adequate to provide them with food, a place to rear their young, and sufficient cover for their protection. A territory once acquired is respected by neighbouring animals of the same species, and trespassers, when attacked, do not defend themselves but retire. Territories are marked by various means. Birds proclaim their dominance over a territory in their song. Mammals mostly mark their territories by means of scent produced from scent glands or in the urine as with wolves and dogs. Animals with a social organisation, as say with monkeys and apes, carry the process a stage further in the 'social hierarchy' system. This is the system of the peck order; in which both males and females arrange themselves in a social order of dominants and subordinates. Amongst horses, the subordinate males are driven from the herd or killed. Amongst baboons and chimpanzees, they are tolerated but the dominant male is usually the one to cover the females. In any case, if numbers become too great, the subordinates are driven away and become easy prey for predators.

In the jungle code, indiscriminate killing of prey animals by predators does not *normally* occur, though overkills are sometimes reported. Predators take what they require and leave the main herd unmolested. The animals mostly taken are the young, which are produced in excessive numbers in any case, and the old and the sick which are more easily caught. The removal of the latter will lead to improvement of the herd generally and is not undesirable. Human beings too, such as the Red Indians and the Eskimos, obeyed these jungle laws, and although they lived by hunting the bison and caribou, never made harmful inroads on the stock, possibly because before the advent of the rifle they lacked the means to do so. They lived in harmony with nature for many generations before the appearance of Western man, the destroyer. The wanton killing of wildlife by Western man, a destruction which in some areas still continues, is an indelible blot on the human record. This is often done because of pure wantonness; at other times it is because of fancied harm done by the animals attacked, such as the usually amiable wolves or the seals which are supposed to deplete fish stocks. This is another instance in which ecological studies are required to lay old bogies.

Ecology, then, teaches us how to live together in well-balanced communities, with benefit to the habitat. Ignorance or disregard of its teachings leads to habitat deterioration or destruction. A general principle of ecology is that the species in a community are 'habitat specific' and in general confined to their own habitat, in which they fit into the community and do no harm. Of course, some species are more habitat specific than others. Those that are too habitat specific, like the Koala Bear, are doomed to extinction when the habitat is altered by natural causes, such as climatic variation. In the long run, the most successful species are those that are not habitat specific and can adapt themselves to different conditions and to different types of food and methods of feeding. Man is such an animal and has carried his nonconformity to new lengths. He not only adapts himself to a wide variety of habitats, but in various ways adapts the habitats to his needs. To interfere with habitats, naturally developed over thousands, even millions of years, is a dangerous practice. Instances abound in which formerly fertile areas have been destroyed in Southern Europe, Africa, India, Pakistan, and America. Now inland waters, estuaries, and even the oceans are threatened. It is, of course, absurd to suppose that the clock can be turned back and that man will again submit to natural habitats. However, with proper study human activities can be correlated with the requirements of the habitat; indeed many habitats could be improved and become more productive.

At the beginning of the chapter, we studied the succession of an ecological sere, by which pioneer surfaces were converted by life successions to a fertile climax. In a sense, earth history can be regarded as such a sere developed from bare rock surfaces and water to the green and pleasant land we know. It will be relevant to our argument and instructive to study earth history in this sense, and this we attempt to do in the next chapter.

NOTES

1 A. Macfadyen, *Animal Ecology: Aims and Methods,* Pitman, London, 1963
2 I.N. Healey, 'The Habitat, the Community and the Niche', in R.N.T -W - Fiennes, *Biology of Nutrition,* Pergamon Press, Oxford, 1972
3 J.L. Cloudsley-Thompson, 'Recent Expansion of the Sahara', *Intern. J. Environmental Studies,* 2, pp.35-9 (1971)
4 J.L. Cloudsley-Thompson and Anne Cloudsley-Thompson, 'Prospects for Arid Lands, *New Scientist,* Nov. 1970, pp.286-9
5 A.G. Tansley, 'The Use and Misuse of Vegetational Terms and Concepts', *Ecology,* 16, pp.284-307 (1935)
6 A.L. Penrose, *Biology of Mental Defect,* Sidgwick & Jackson, London, 3rd ed., 1963
7 C.S. Elton, *Voles, Mice and Lemmings,* Oxford University Press, London, 1942

Further Reading

Allen, W.C., Emerson, A.E., Park, O., Park, T., and Schmidt, K.P., *Principles of Animal Ecology,* Saunders, Philadelphia and London, 1949

Elton, C.S., *The Ecology of Invasions by Animals and Plants,* Methuen, London, 1958

Elton, C.S., *The Pattern of Animal Communities,* Methuen, London and New York, 1966

Errington, P.L., 'Factors limiting Vertebrate Populations', *Science,* pp.124-304, 1956

Fiennes, R.N.T -W -, *Man, Nature and Disease,* Weidenfeld & Nicolson, London, 1964

Roberts, D.F. and Harrison, G.A. (eds.), *Natural Selection in Human Populations,* Pergamon Press, Oxford, 1959

2 THE FORMATION OF THE EARTH AND THE ORIGINS OF LIFE

The Origins of Life

The probiotic earth – that is the earth before life existed on it – differed from the earth of today in two important features. First, although there was a gaseous atmosphere, this was devoid of free oxygen. Secondly, and linked with the first, the ozone layer of the upper atmosphere was absent. Since the ozone layer traps most of the sun's ultra-violet radiation, the earth's surface was exposed to a degree of radiation which would be lethal to any form of life that exists today unless protected, as beneath water.

It is generally accepted that the earth started as a cold body formed by aggregation and concretion of spatial debris, which originated in the sun.[1] Internal heat was generated by pressure and other forces, by which deep rock formations became molten, leading to intense volcanic activity. Exhalations from this activity led to the formation of an atmosphere in which hydrogen, nitrogen and carbon gases predominated; water vapour was present but no free oxygen. Their low concentration in the atmosphere today compared with their cosmic abundance argues that the lighter gases, such as neon and helium, escaped into space from the earth's gravitational forces.

The earth's surface was under constant bombardment from meteorites, which could not burn out in an atmosphere devoid of oxygen. Their impact created pitting and cratering, adding to the effect of volcanic activity. The surface was, therefore, by no means uniform in structure, nor was it level. In depressions and craters, often very large, water accumulated and eventually seas formed. Plainly, too, climatic differences existed between the poles and the equator, contributing to the lack of uniformity, and there were great temperature differences between night and day.

Table 1 *The Origins of Life*[2]

	(Years ago)
Origin of the earth	4,500 – 5,000 million
First traces of life (bacteria)	3,300 million
Simple organisms containing chlorophyll	2,700 million
Metazoa (multi-cellular life)	600 million
Life on land	400 – 500 million
Dominance of mammals and end of age of reptiles	100 million
Appearance of man	200,000
Appearance of modern man	20,000

The earth's probiotic era lasted 1,200 million years. There was no oxygen in the atmosphere, no life, fierce ultra-violet radiation, bombardment with meteorites, and abundant volcanic activity.[3] There were rocks and seas, and the elements we know today were present in a disordered state. Yet, in the sun's radiation, there was a far more powerful energy source than is available today because of admission of the ultra-violet radiation. It is known from experiments in laboratories that such radiation can cause the combination of organic chemicals, of which the most important are oxygen, hydrogen, carbon and nitrogen, to form the so-called 'biochemicals' which are the basis of life, namely amino acids, polypeptides, proteoses and proteins. From this it is postulated that, in suitable areas and under suitable conditions, there must have accumulated masses of these biochemicals in a so-called 'primordial soup'. In the depth of this primordial soup and under conditions of protection from lethal ultra-violet rays, the random meeting of diverse biochemicals could result in the evolution of some form of primitive self-replicating chemical system and so to a living thing. The power source of an early life form could not be the sun's rays, so that it would live saprophytically on the 'primordial soup' from which it was developed. There are no real improbabilities or perplexities in this concept. Possibly, the only real difference between a self-replicating chemical system and a living system is that the latter is enclosed in a cell wall or pellicle. Such could develop initially by a condensation or drying of the outer portion of the soup around the chemical system.[4]

It is reasonably well established that such could have happened under the conditions prevailing. There is no proof that it did happen. Life could conceivably have reached earth from outer space in meteorites or even in space ships or flying saucers and found a suitable medium to propagate in the primordial soup. Two important points are evident. First, the earliest life forms could not have used energy systems requiring oxygen. Secondly, they must have lived saprophytically on pre-existing chemicals – a strong argument in favour of primordial soup. In addition, they must have been protected from the ultra-violet rays of sunlight, and it is argued as most probable that this would have occurred some five metres below the surface of water, where such protection would exist. This would explain the existence of the earliest forms of recognisable life in the seas.

Creation of the Biochemicals

Although oxygen was absent from the atmosphere, it was abundant on the earth, chemically bound in the rocks and combined with hydrogen in the waters of the seas. This oxygen was gradually released, chiefly from the waters, by photo-dissociation. However its accumulation was limited by a natural process, known as the Urey effect, to approximately 0.1 per cent of today's atmospheric pressure, because the oxygen produced by photo-dissociation shaded the underlying water vapour from ultra-violet radiation. In addition to molecular oxygen, chemically unsatisfied forms of oxygen (0 and 03-ozone) were produced near the earth's surface and the ozone formed a thin layer above the molecular oxygen. Oxygen itself as well as ultra-violet light was toxic to the protoplasm of early life forms, which were not metabolically adapted to use it. The first oxygen to appear on earth was thus a pollutant. When plant life appeared, this pollution was intensified, because the oxygen excreted into the atmosphere was increased by photosynthesis. The development of photosynthesis was one of the great turning points in earth history; to it the earth owes the oxygen of its atmosphere, the raising of the ozone layer, and the development of higher life forms dependent on oxygen for their metabolic processes.[5]

The earliest earth rocks available for study are the Pre-Cambrian of some 600 million years ago, at which time metazoan life was already present. Estimates of the age of the oldest moon rocks so far acquired give a figure of 4.6 million years, and this accords well with the estimate given of earth's age of 4,500 to 5,000 million years. There are, however, no rocks of comparable age on earth which could give a clue to its composition at that time. This is not surprising since alterations due to weathering by the elements have continuously occurred on earth, but are absent from the moon. On the other hand, the moon's gravitational field is inadequate to retain more than a fraction of the gases originally present, so that the moon's atmosphere can give no clue to the atmosphere likely to have been present on earth at that time. Whether of similar origins or not, it is evident that very different influences have affected the earth's surface from those which affected the moon's. Whereas the moon's surface has altered little in the space of 3,000 million years, the earth's has been completely metamorphosed. However, on earth the release of gases from primordial rocks led to certain elements of stability which have persisted to this day, such as the concentration of salt (chloride ions) in the sea, the pH of sea water and the carbon dioxide content of the atmosphere.[6]

The situation in relation to carbon is one that bears more detailed examination. The quantities of industrial carbon dioxide reaching the atmosphere today are far in excess of what must have reached it in probiotic times from volcanic exhalations and other sources. The additional quantities are in excess of what could be removed by plant photosynthesis. Yet, any rise in atmospheric carbon dioxide is marginal only, because of the absorptive capacity of the oceans. So great is this, indeed, that it is not even reflected in a measurable alteration in the pH of sea water.

The question of the atmospheric gases has been discussed by N. W. Pirie.[7] Pirie points out that the atmosphere now contains 12 x 10^{20} g of oxygen, which is 235 g/cm^2 of the earth's surface. Of fossil oxygen, consumed in making sulphate and ferric iron, there may be from 1200–1500 g/cm^2. However, significantly more oxygen must have passed through the earth's atmosphere than is now present in it, so that the total of present and past oxygen may be in the region of 1800 g/cm^2. If most of the oxygen that has passed through the atmosphere is derived from photosynthesis, there should be an equivalence between oxygen and carbon now present in reduced forms, such as coal and oil. Such an equivalence cannot be shown, unless account is taken of kerogen deposits, that is the reduced carbonaceous material in sedimentary deposits. Though the figure reached in this way at 2,000 g/cm^2 is rather too high, when adjustment is made for carbon released from methane, carbon monoxide and so on, the equivalence is very close. Another estimate reveals that the present-day plant cover could make the amount of oxygen now present in the atmosphere in two to three thousand years. Thus once chlorophyllous plants had colonised the earth, the stable atmospheric system must have been developed with great rapidity; only the saprophytes lagged, resulting in locking away a great deal of carbon but without much effect on atmospheric stability.

The third important gas in the atmosphere, with oxygen and carbon dioxide, is nitrogen. It is generally held that nitrogen was present in the primitive atmosphere in the form of ammonia; this would be analogous to the ammonia-rich atmosphere of the giant planets. The source would be volcanic gases, there being a certain amount of ammonia present in plutonic rocks. However, since nitrogen is of approximately the same molecular weight as the rare gases neon and argon, it is virtually certain to have been lost into space with them. The nitrogen in the present-day atmosphere is, therefore, most likely due to biological activity. We shall see in a later chapter how the atmospheric levels of nitrogen, oxygen and carbon dioxide are kept stable by precisely regulated cycles.

We must now take a further look at the primordial soup and the ways in which living matter might have developed in it. This will involve also a consideration of the nebulous dividing line between life and non-life. As we have seen, the earth was already over 1,000 million years old when the first traces of life became apparent. The first recognisable living organisms were bacteria. Thus, even if more primitive life forms existed before the bacteria, in all this time life had not advanced far along the evolutionary scale. During this time, provided that the basic organic chemicals were assembled together under suitable conditions, the more complex 'biochemicals', consisting of the larger molecules and polymers of carbon, nitrogen, hydrogen and oxygen, could have been synthesised by the power source of solar radiation. Accumulating over this vast period, almost every conceivable variation of chemical reaction could have occurred. Amongst the chemicals likely to have appeared would be self-replicating structures and deoxyribonucleic acid (DNA), two strands of which could by a random fit have formed a double helix with appropriate linkages. The same random process would in the course of time bring together the other essentials of life, such as the organic catalysts, the energy transformers (adenosine diphosphate (ADP) and adenosine triphosphate (ATP), and the higher polysaccharides. Brought together randomly, an organised system having the characteristics of life could have appeared, and, given the conditions believed to have been prevailing during this immense span of time, surely must have appeared.[8]

Most students of the subject believe that this is what did occur. If so, the most primitive life forms obtained their energy sources from biochemicals already existing in the primordial soup, using fermentation processes such as those of yeasts which break down glucose molecules into alcohols. The use of such energy sources requires that an otherwise stable molecule, such as that of glucose, be degraded under catalytic action; the primitive organisms, therefore, needed to produce the necessary system of catalysts, a DNA system to reproduce themselves and their catalysts, and an ATP system to direct energy to constructive channels. One further property is required of a living system, namely that the stable, self-replicating, package once developed be enclosed in a membrane, without which dispersion would occur. Cell membranes are in themselves complex structures with active metabolic properties. First, they must be physically protective of the package inside. Secondly they must admit water and nutrients selectively, while excluding unwanted substances. Thirdly, they must be able to excrete waste products. Bacteria themselves possess an outer cell wall which is inert, and an inner cell membrane which is metabolically active in these ways; that is to say that the cell membrane draws on the power sources of the cell to admit and discard materials in an organised way.

Evolution of the Cell

The evolution of the functional cell membrane must have occurred over a long period of time during which the living entity survived in a medium whose composition was similar to that within the cell itself. Energy sources in those early times were not confined to the breakdown of only one type of chemical. One supposes myriads of alternatives have been available. Amongst primitive organisms which survive to the present, very many different chemical reactions are used as power sources. This has been mentioned in Chapter 1 and will be further discussed in a later chapter. All the evidence suggests that very many metabolic systems were tried and discarded, and that those found most

Fig 1 THE BACTERIAL CELL. Section of a bacterium as seen in electron micrographs. W cell wall, P plasma membrane, R ribosomes embedded in a protein matrix, D threads of DNA. Note absence of nuclear membrane.

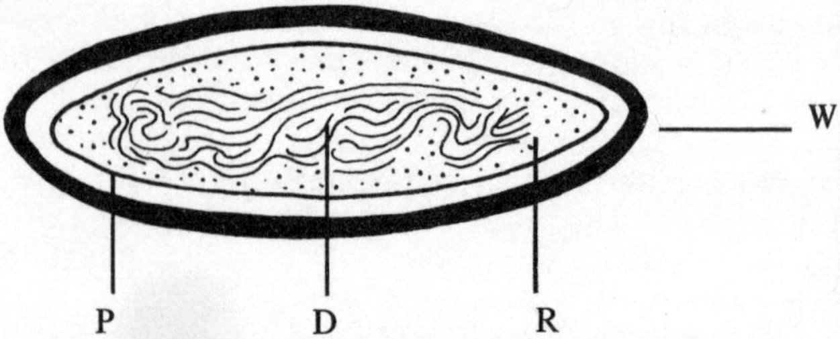

Fig 2 THE EUKARYOTE CELL. The primary divisions of the cell: nucleus, cytoplasm and plasma membrane.

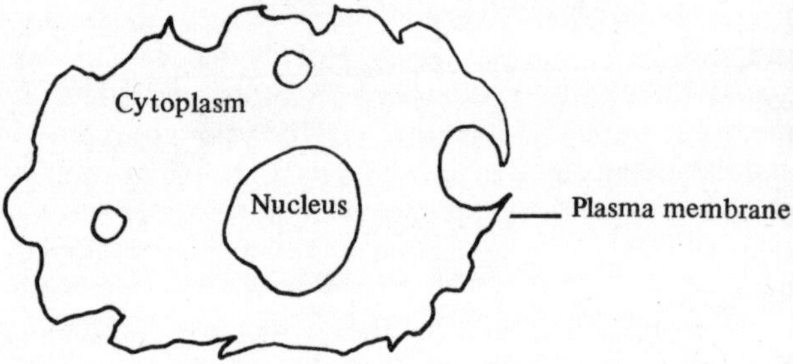

Fig 3 THE EUKARYOTE CELL: FINE STRUCTURE. Cytoplasmic structures of a cell of generalised structure as seen in electron micrographs N. nucleus, NM nuclear membrane, O nucleolus, G Golgi apparatus, M mitochondria, R reticulum of particle-covered membranes (endoplasmic reticular), P ribosomes.

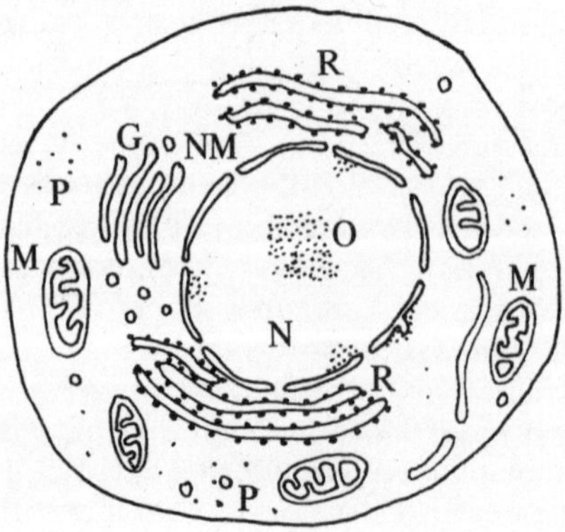

Fig 4 THE EUKARYOTE CELL: ACTIVE AND INACTIVE. The cytology of two forms of cells synthesising proteins: (a) glandular, secreting cell. N nucleus, G Golgi apparatus, S secretory granules. Note numerous particle-covered membranes and polarisation of cytoplasm; (b) retaining cell. M mitochondrion. Fewer Golgi membranes, no particle-covered membranes. Numerous ribosomes.

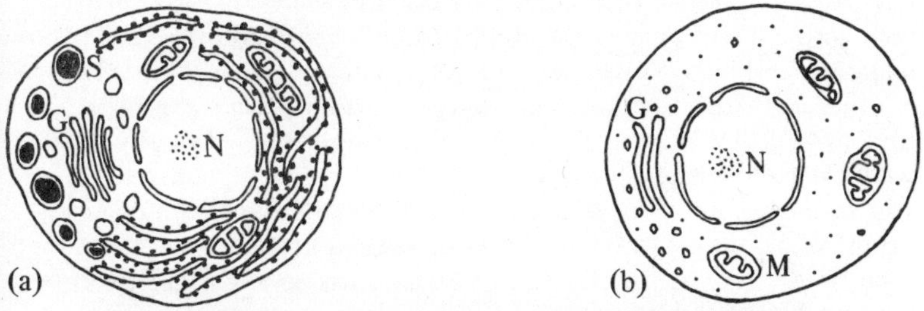

Fig 5 THE EUKARYOTE PLANT CELL. Fine structure of a meristematic (embryonic) plant cell. The nucleus contains a large granular nucleolus n and is enclosed in a double-layered membrane pierced by 'pores'. The Golgi cluster of membranes G is well developed, and several paired membranes D, perhaps early steps in the formation of vacuoles such as V, are shown (m are mitochondria). The cell wall is usually thin at this stage. The cytoplasm is very dense, with ribosomal particles as in embryonic animal cells. PM is plasma membrane.

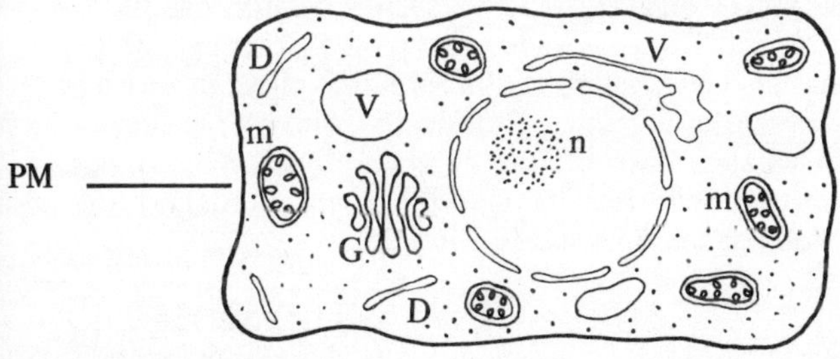

EUKARYOTE CELLS

There are two primary subdivisions of eukaryote cells, nucleus and cytoplasm. The nucleus contains the genetic material and the code, which directs the cell's activities. The cytoplasm organises those activities and, through the mitochondria, directs the cell's energy exchanges. The various organelles present in the cytoplasm are important for these functions and are seen to be more developed and more active in secreting cells than in structural cells.

Compare the much simpler prokaryote cell of the bacterium, which can nevertheless manufacture a wide range of enzymes.

advantageous were retained. Thus, as with life in general and in conformity with the principles of ecology, randomness gave way to order.

The bacterial cell has no nucleus, the DNA being dispersed throughout the cell cytoplasm. There are genes controlling heredity, but these are not organised as chromosomes. Most experts agree that they have no organised mitochondria which form the powerhouse of nucleated organisms. It took another 600 million years for a eukaryote (cell with a nucleus) to be developed from the prokaryotes (cells without a nucleus). The development of the eukaryotes must be regarded as another of the outstanding milestones in earth's history, especially since, when they appeared, many of them contained chlorophyll and were capable of true photosynthesis. However, chlorophyll was not a new invention since it is present in the green and purple bacteria, very primitive organisms which live in estuaries and use sulphur sources for their energy requirements. In addition they use energy acquired from light at the red end of the spectrum, but they do not synthesise carbohydrates from carbon dioxide, nor do they excrete oxygen.[9]

The blue-green algae were the first chlorophyll-containing organisms to appear which could synthesise carbohydrates, evolving oxygen into the atmosphere during the process. Like some fungi, they were coenocytes, that is to say that though they were eukaryotes the nuclei were not separated into individual cells but there were many nuclei in the cytoplasm of one cell envelope. The chlorophyll too was distributed throughout the cell sap and not confined to chloroplasts or plastids, as in more highly evolved organisms. Some, such as *Nostoc*, can fix atmospheric nitrogen. The appearance of the blue-green algae was followed by that of the unicellular chlorophyllous algae, in which the chlorophyll is present in organised plastids. In was from the organisation of these one-celled algae into colonies, with eventual differentiation of function amongst the cells, that the metazoa or multi-cellular life forms developed. To look at the time scale again:

	(Years ago)
Origin of the earth	4,500 – 5,000 million
Bacteria	3,300 million
Blue-green algae	2,700 million
Fungi	2,300 million
Metazoa	600 million

It has, thus, taken the greater part of earth's history, 4,000 of say 4,600 million years, to produce a simple animal like a jellyfish.

Before passing to the pollution of earth's atmosphere with oxygen, let us take a quick look at the great evolutionary step resulting in the appearance of the eukaryote cell. The first eukaryotes to appear in the earth's record were the blue-green algae and the fungi, both coenocytes though some fungi and some stages of fungal development are unicellular. The fungi, like the bacteria, live saprophytically and use a multiplicity of chemical processes to obtain their energy requirements. The unicellular organisms, protozoa and protophyta, followed, and with them was developed a sophisticated type of cell, in all respects comparable with the cells that are the basis of metazoan life from worm to man. Each cell contains a complete metabolic system, is equipped for locomotion, to detect and seek food, for ingestion absorption and excretion, for sexual and asexual reproduction, for gene inheritance, and for power utilisation. It is sometimes suggested that these cells, and all cells of metazoa, started as hybrids in the sense that they are not uniform structures. This idea started with the plastid, the structure which contains chlorophyll and closely resembles some free-living protozoa. It is supposed that the plastid may have originated from a protozoan which came to live in symbiosis with a non-chlorophyll-bearing protozoan of another type, and gave it protection and powers of mobility. Such an idea has support, because in the early days of eukaryote evolution before oxygen was plentiful in the atmosphere, symbiosis became a feature of a number of primitive organisms, such as sea anemones, many of which harbour green flagellates in their tissues. Secondly, when cell division takes place, the plastid divides independently of the host cell. Thirdly, plastids have been removed from the host cells and can for a certain time live independently of them, carrying on their work of photosynthesis.

It has also been suggested that the cell and its nucleus may originally have been separate entities which came together fortuitously. For instance, the nucleus of one cell can be removed and transplanted into another cell similarly deprived of its nucleus; it can take control and regulate the functions of the cell into which it is transplanted. These are interesting speculations for which there can be no proof, but which could explain the ways in which the eukaryotes were evolved. In any case, primitive eukaryotes have as diverse methods of acquiring and using energy as have more primitive life forms. It was not until much later that energy use became standardised in oxidation/reduction processes; indeed this may have been a process forced on living systems outside specialised habitats by the increasing pollution of the atmosphere with oxygen.

Once the blue-green algae and the unicellular chlorophyll-bearing protozoa appeared on earth, a new driving force of life came into existence dependent on photosynthesis, that is the trapping of the sun's protons on chlorophyll and use of the energy to convert carbon dioxide into carbohydrates and fat in which the energy is stored in chemical form for recovery when required. It suffices here to say that the raw material of the chemical process, carbon dioxide, is acquired from the atmosphere, and oxygen is the end product which is excreted into the atmosphere. The earlier life forms, as we have seen, required pre-existing molecules of organic materials or biochemicals on which to live. These would become scarcer as some screening of the earth's surface occurred by oxygen and ozone derived from photo-dissociation of water and plutonic rocks, and in any case would not occur at any depth in water. It is unlikely that saprophytic life could have continued indefinitely unless a new energy source could be tapped. Such now became available, but owing to the poisoning of the atmosphere by oxygen it would also have become self-limiting unless new metabolic mechanisms had been evolved, to which the presence of oxygen would be inimical. Furthermore, when the oxygen pressure reached 1.0 per cent of present levels – known as the Pasteur point – oxidation/reduction systems of energy use become more efficient than fermentation. This point was reached at the beginning of the Cambrian Period some 550 million years ago, when a great population explosion of varied life forms occurred. This was another of the great landmarks in earth history. Atmospheric oxygen must have been built up steadily from then on, until the stable condition of earth's environment of today was reached. In time the ozone belt was lifted to a sufficient height to enable colonisation of the land as well as the sea, and the age of lichens, mosses, giant ferns, and gymnosperms led into the Carboniferous era responsible for our fossil fuels.

Since that time, the earth has passed through many vicissitudes. Some aspects of this history we shall study in greater detail in later chapters. The great and prolonged age of reptiles came and went. The primitive and originally stupid warm-blooded mammals became dominant and radiated into the world's habitats, evolving into many forms, extinct and extant today. The whole proceeded according to the dictates of ecology, until man came to defy ecology some 10,000 years ago, and that must be regarded as the last of our turning points in earth history. As a final exercise in this chapter, we can list earth's turning points:

	(Years ago)
Birth of the earth	4,500 – 5,000 million
Appearance of first life (bacteria)	3,300 million
Photosynthesis	2,700 million
Eukaryotes (fungi)	2,300 – 1,000 million
Metazoa	600 million
Pasteur point	550 million
Agricultural revolution of man	10,000 million

In nearly 5,000 million years of earth's history, there have been but six revolutionary turning points. Man's interference with natural ecological processes, only 10,000 years ago, has been one of them.

NOTES

1 L.V. Berkner and L.C. Marshall, 'History of Major Atmospheric Components', *Proc. Nat. Acad. Sci.* vol. 5, 53, pp. 1215-26 (1965)
2 E.C. Olsson, 'Climatic Change and its Influence on Life and Habitat', in R.N.T -W - Fiennes, *'Biology of Nutrition'*, Pergamon Press, Oxford, 1972
3 Berkner and Marshall, op. cit.
4 For a review of chemical evolution, see Crawford, in Fiennes, op. cit.
5 For reviews, see Berkner and Marshall, op. cit., and L.V. Berkner, 'On the Origin and Rise of Oxygen Concentration on the Earth's Atmosphere', *J. Atmos. Sci.*, 23, pp.133-43 (1965)
6 Berkner and Marshall, op. cit
7 N.W. Pirie, 'Characteristics of Living Things', in Fiennes, op. cit.
8 Crawford, op. cit.
9 See E.J. Ferguson-Wood, 'The Food Cycle in Marine Environments: Marine Algae', in Fiennes, op. cit.

Further Reading

Bernal, J.D., *The Origin of Life,* Weidenfeld & Nicolson, London, 1967
Fiennes, R.N.T -W - , *Biology of Nutrition,* Pergamon Press, Oxford, 1972
Fox, S.W., *The Origin of Prebiological Systems,* Academic Press, N.Y., 1965
Kulp, J.L., 'Geological Time Scale', *Science,* 133, pp.1105-14 (1961)
National Academy of Sciences Symposium, 'The Evolution of the Earth's Atmosphere', *Proc. Nat. Acad. Sci.* (Wash.), 53, p.1169
Oparin, A., *The Origin of Life on Earth,* Oliver & Boyd, Edinburgh, 3rd ed., 1957
Oparin, A. (ed.) *The Origin of Life on Earth,* Pergamon Press, Oxford, 1964
Oparin, A., *Life: Its Nature, Origin, and Development,* Oliver & Boyd, Edinburgh, 1961
Stanton Hicks, Sir C., 'The Nutritional Requirements of Living Things', in Fiennes, op.cit.

3 PRIMITIVE LIFE FORMS AND THE EVOLUTION OF METAZOA

The metazoa, which we regard as the higher forms of life, could not exist without the lower forms evolved in the early days of earth history. The lower forms, on the other hand, could, and for 2,700 million years did, exist without the higher. Furthermore, their biomass, their total weight of living material, vastly exceeds the total biomass of the vertebrate kingdom. It is well, therefore, to devote a chapter to their study and to examine their activities.

Whereas the metazoa, whether plant or animal, employ a single energy system, the reduction/oxidation or 'redox' system, lower organisms employ a variety of methods to obtain and use their energy. At this point, it will be helpful to introduce a few terms that are used to describe methods of nutrition. Living things are either 'autotrophs' or 'heterotrophs'. *Autotrophs* use inorganic nutrients, and fix carbon dioxide or use carbonate or sodium bicarbonate as their source of carbon. This group thus includes the plant kingdom. *Heterotrophs* require organic compounds both for energy metabolism and carbon assimilation. The group thus includes the animal kingdom. Living forms which use stored resources for their energy requirements, as do the metazoa, are described as *endergonic;* those that use external energy sources are known as *exergonic.* Yeasts are typical examples of exergonic organisms, since they obtain their energy requirements by degrading glucose, present in their environment, into alcohols. *Chemotrophs* are organisms obtaining energy by the oxidation of chemical compounds derived from the environment (exergonic). Chemotrophs are of two groups: *chemolithotrophs,* which use oxidisable inorganic substances, such as molecular sulphur or hydrogen sulphide or ammonia; and *chemo-organotrophs*, which use only organic substances for the same purpose.

Phototrophs obtain their energy from radiant energy. They contain chlorophyll and are able to trap and use the photons of incident light.

Photo-lithotrophs use a combination of radiant energy and oxidisable inorganic compounds.

Photo-organotrophs use a combination of radiant energy and oxidisable organic compounds.

Two other terms are worthy of mention: *phagotrophy* or *phagocytosis,* which is the ingestion by single cells of other cells or particles; and *pinocytosis,* the ingestion by individual cells or droplets.

Within these definitions, there exists an enormously complex variety of mechanisms for feeding and energy use. The higher animals of today

Fig 6 ENERGY PATHS OF AN ECOSYSTEM 1. Fate of energy incorporated by autotrophs in Cedar Bog Lake, Minnesota in gram calories per square centimetre per year.

Data of R Lindeman. 1942. Ecology 23: 399-418

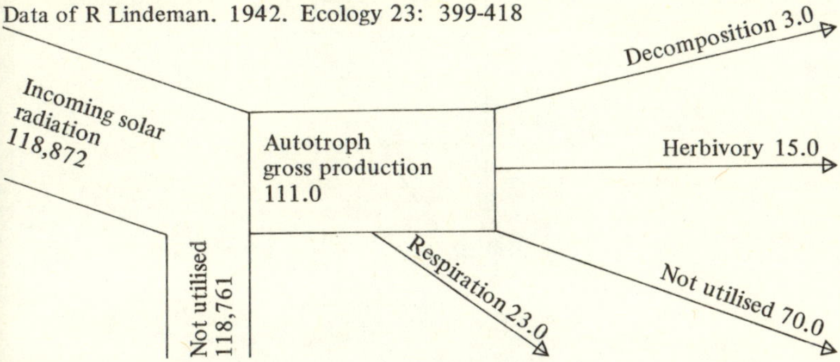

Fig 7 ENERGY PATHS OF AN ECOSYSTEM 2. Fate of energy incorporated by herbivores in Cedar Bog Lake, Minnesota in gram calories per square centimetre per year.

Data of R Lindeman. 1942. Ecology 23: 399-418

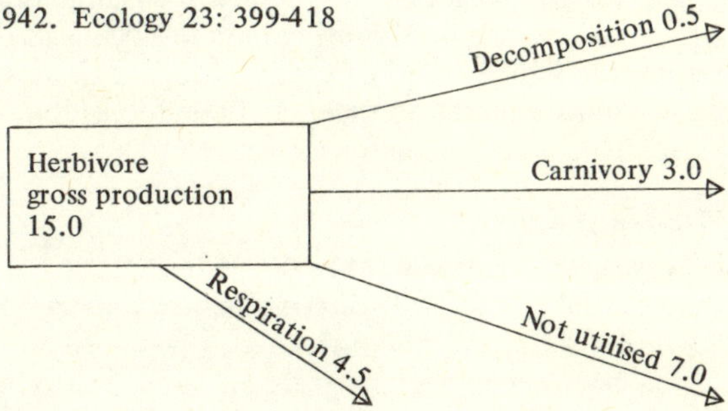

Fig 8 ENERGY PATHS OF AN ECOSYSTEM 3. Fate of energy incorporated by carnivores in Cedar Bog Lake, Minnesota in gram calories per square centimetre per year.

Data of R Lindeman. 1942. Ecology 23: 399-418

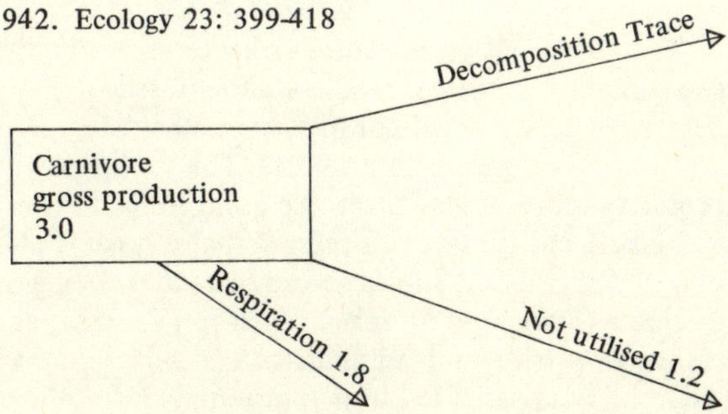

Fig 9 ENERGY PATHS OF AN ECOSYSTEM 4

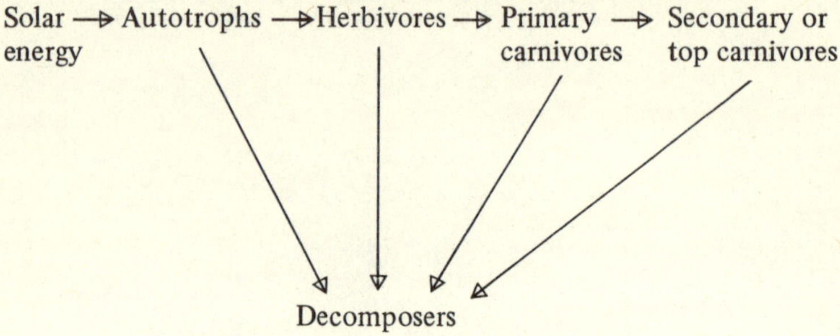

have virtually lost the power to digest vegetable matter; they are unable on their own account to split the molecules of cellulose and other higher polysaccharides. Thus, herbivorous animals possess large fermentation chambers, in which micro-organisms do the job for them. Furthermore, higher animals are no longer able to synthesise some of the vitamins essential to them, and this too is done for them by bowel-living micro-organisms.

Micro-organisms

However, we must now study the various groups of micro-organisms that are essential components of ecosystems. These are the bacteria, the fungi, the algae, and the protozoa.[1]

The bacteria, as we have seen, are supposed to have been the earliest forms of true life to have been evolved. These organisms possess a multitude of ways in which they obtain their energy. They are, of course, responsible for many infectious diseases, but their activities are for the most part beneficial and our kind of life could not continue without them.

The green and purple sulphur bacteria are the only bacteria to contain chlorophyll, as we have already seen. They are found in habitats where few other life forms can exist, that is among the rotting detritus of sea grasses in estuaries. The decomposition of the detritus produces more or less anaerobic conditions, that is absence of oxygen, and great acidity. For this reason, many organic chemicals are present in reduced (more complex) form instead of being oxidised; there is methane (marsh gas) in place of carbon dioxide, ammonia instead of nitrate or nitrite, and sulphide instead of sulphates. There are also a great many unsatisified iron compounds, often linked with sulphur.

Stagnant lagoons become black with iron compounds and the smell of hydrogen sulphide becomes overpowering. Such conditions are suitable for the green and purple sulphur bacteria, which are virtually unique in being able to live in such a situation. These bacteria are photosynthetic, but obtain their energy by knocking electrons off H_2S (hydrogen sulphide) instead of H_2O (water), used by other photosynthetic plants; this is the reason why their photosynthetic activities do not release oxygen. The ecology of such a habitat depends for its initiation on the photosynthetic activities of these bacteria.

When conditions become less extreme, some other bacteria continue the process. Ciliates (protozoa) prey on them, and even some nematode worms, crustaceans and molluscs become established. The sea grass flats which then appear are very unproductive habitats, but with understanding of the ecology and proper treatment, large harvests of fish protein can be produced by culture methods. The coloured bacteria survive beneath the other vegetation, because they only use light toward the red end of the spectrum. They can thus 'scavenge' the more penetrating wavelengths, which have filtered through carpets of sea grasses.

Some chemolithotrophs (sulphur bacteria) also use sulphur and are abnormally resistant to acid media. One indeed can tolerate strong sulphuric acid, and some of them can cause erosion of concrete and alkaline rocks. The important soil-nitrifying bacteria belong to this group, converting ammonia to nitrite and then nitrate ($NH_4 \rightarrow NO_2 \rightarrow NO_3$). Another group uses molecular hydrogen as an energy source, and another group iron compounds. The latter occur in iron-containing soils such as bogs and swamps, where they produce rust-coloured patches. Where present in some forest soils, they may precipitate iron below the topsoil causing an iron pan to develop and converting the forest in time into bog.

Most true bacteria (non-chlorophyllous) and all the fungi are chemo-organotrophs. Metabolically, bacteria and fungi are rather similar and the fungi probably evolved from the bacteria. Although the bacteria are mostly unicellular, there are important filamentous forms which appear to be intermediate, such as *Actinomyces.* However, all fungi have primitive eukaryote nuclei, whereas the filamentous bacteria do not; nor do bacteria have distinct mitochondria, which fungi have. The more primitive fungi such as the yeasts have rather primitive nuclei and they appear to be the earliest representatives of eukaryotes. Both bacteria and fungi are variable in their requirements of oxygen. Some can only live in an atmosphere containing oxygen; some can only live in an atmosphere devoid of oxygen; yet others require diminished oxygen; and others finally can live either in an aerobic or an anaerobic atmosphere. Their metabolic pathways can be oxidative or fermentative.

The oxidative reaction requires molecular oxygen as the final electron acceptor. The fermentative reaction is an incomplete one in that molecular oxygen is not required and energy requirements are obtained from organic chemicals only degraded part-way to products of lower molecular weight. In both cases, the energy is transferred to the substance ATP, described later.

The range of materials that can be attacked by the chemo-organotrophic bacteria covers virtually any organic materials found in nature. It is, however, an anomaly, revealing of the hit-and-miss nature of chemical evolution, that there is a great dearth of organisms capable of attacking the man-made synthetic organic chemicals, whether the organo-chlorine agricultural chemicals or the plastics, such as polythene and polystyrene. For this reason, the man-made synthetics have become a serious problem of waste disposal and pollution. This immunity from attack, in another way, has its uses because these substances are not only spared microbial action, but are not even attacked by the defence cells and fluids of the vertebrate body. They can, therefore, be used for organ implants, such as heart valves, or teeth, and will not be resorbed.

Amongst the hydrocarbons, there is one bacterial genus, *Methanonomas,* which can oxidise methane, and many can to a certain extent use ethane, propane and butane. Some yeasts too can attack a range of such hydrocarbons. There are even organisms capable of attacking fractions of crude oil, and it has been suggested that by their use food products could be synthesised from crude oil.

There exist species of bacteria and fungi which can split virtually all the higher polysaccharides of high molecular weight, some of them producing intermediate products, such as dextran, which are of commercial value. Organisms with such properties are also responsible for 'dry rot' of timber, 'soft rot' of vegetables and other deleterious processes. Organic acids are commonly released during the fermentation of carbohydrates; acetic acid is made by the vinegar bacteria; propionic acid important in the manufacture of soft cheeses is made by *Propionibacterium;* and lactic acid, responsible for clotting of milk, is made by lactobacillus. Some organic acids are themselves utilised as energy sources; such are citrate, tartrate, malonate and lactate.

Other organisms obtain their energy and carbon from breakdown of amino acids, by removal of the amino group. Others produce enyzmes capable of attacking nucleic acids. Fats too are hydrolysed by some groups of bacteria to yield glycerol and fatty acids. There are even bacteria which can degrade chitin, the material of the hard shells of lobsters and insects.

Bacteria and fungi then are the great scavengers of the world. Man-made synthetics apart, there are few organic materials they cannot decompose. In this, they play a vital part in the economy of nature.

We must revert now to the blue-green algae already noted as the earliest true plants in which the full photosynthetic process had been developed. It was due to their influence that the atmosphere first became seriously contaminated with oxygen. Structurally, the blue-green algae are primitive and show greater affinities with the green bacteria than with the true algae. Like the bacteria, they do not have a defined nucleus within a nuclear membrane; their chlorophyll is dispersed throughout the cell body, not in defined chloroplasts or plastids; and like the bacteria, they do not show the streaming of cytoplasm characteristic of higher forms, such as fungi. On the other hand, they are mostly filamentous; they possess chlorophylls resembling those of true algae and higher plants; and they use carbon dioxide and water in the process of photosynthesis instead of hydrogen sulphide, used by the green bacteria. Like fungi, their reserve food is glycogen (animal starch), not starch. The filaments of the blue-green algae are embedded in a tough mucilaginous sheath, so that the colony becomes a slimy green layer often seen in river estuaries at low tide. Underneath the slime are often also found the green bacteria. They may also be free-floating in water or have filaments radiating into the water. A large group also lives in soil habitats and these have the power, rare in nature, of fixing atmospheric nitrogen as well as using light as an energy source. Some, protected by their mucilage, can even creep over hard substrates. Thus their range of habitats, which are always primitive, is wide, and they have been found both in arctic waters and in hot springs. Some blue-green algae can taint water supplies or render them toxic to animals.

Perhaps the chief importance of the blue-green algae is in maintaining the aeration of estuary waters, a role which is vitiated when there is estuary pollution by sewage; this stimulates the growth of a surplus of oxygen-using micro-organisms, which quickly deplete the waters of oxygen and make it impossible for fish and other forms of aquatic life to survive. As pioneers, some blue-green algae live symbiotically with fungi as lichens. In this role, both their symbiotic and their nitrogen-fixing properties would be of value to the fungus, and the fungus, able to absorb moisture from the atmosphere, would protect the algae from desiccation.

Metabolically, the blue-green algae have taken a great step forward over the green bacteria, in that they have developed a form of true photosynthesis in which water replaces hydrogen sulphide in energy conversion and oxygen is given off into the atmosphere. This revolution paved the way for the evolution of higher life forms. Nevertheless, they still retain an important place in earth's ecology, both in water and on land.

Unicellular eukaryôte plants do not appear in the fossil record until some thousand million years ago, although eukaryote fungi were in existence some two thousand million years ago. It cannot be said, therefore, whether the unicellular plants were developed from fungi which acquired chlorophyll, or from blue-green algae which independently developed a eukaryote nucleus, or from some other form of life of which we have no record. The material, and the sources from which it is obtained, are so scanty and in other ways unsatisfactory, that any conclusions based on them must be largely speculative. Suffice it to say that further evolution, the development of the metazoa and higher life forms, depended entirely on the eukaryote cell. The chief feature of this cell is the specialisation of function within the cell, which is seen in three basic ways. First, the genetic chromatin is packed into a nucleus within a definite nuclear membrane. Secondly, the power systems of the cell are packed into definite organelles, known as mitochondria. Thirdly, the cell chlorophyll is packed into an organ of increasing complexity known as the plastid or chloroplast. Organs of ingestion, excretion, food digestion and storage, and locomotion also appear amongst some single-celled eukaryotes. At the same time, genetic material has become aligned on chromosomes and mating cells undergo reduction division (meiosis) preparing them for fusion.

Since the appearance of both fungal and plant eukaryotes, the fungi have gone their own way, retaining their primitive methods of obtaining energy, and have not contributed to the main evolutionary stream. It is, therefore, to the plant eukaryotes that we must look for further development. Many of these forms, especially the green flagellates, have the ability to live either as autotrophs or as heterotrophs; they can live by photosynthesis alone, or by assimilation of organic materials; alternatively some photosynthetic organisms require organic material as well. In addition, there exist protozoa, otherwise identical with those that are photosynthetic but which have no chloroplast. If some photosynthetic protozoa are kept in the dark with a sufficiency of organic material in the medium, they will lose their chlorophyll but regain it when again exposed to light. Amongst these primitive organisms one cannot therefore distinguish between plants and animals. However, once evolved these eukaryotes developed a balanced ecological system of their own, consisting of plants, plant-living organisms, grazers, predators and scavengers.

Evolution of Metazoa

The beginnings of our modern world came with the development of metazoan life some 600 million years ago, and the great explosion of life forms in the seas with the beginning of the Cambrian period some 550 million years ago. At that time, the foundation stock of all our

37 modern life forms had appeared. Once started, therefore, the colonisation of the earth's habitable surfaces proceeded with comparative speed. The evolution of the metazoa was initially a somewhat simple process, and can be traced in colonial forms that still exist today. In these, the individual flagellate cells are arranged in a sphere with the flagella facing outwards, the whole colony protected by mucilage, as with *Volvox; Pandorina* has the form of a solid sphere, and *Gonium* a flat plate. In some, fine protoplasmic strands connect the cells. They are found in fresh, marine, and brackish waters. Amongst them, division of labour is to be found in the more advanced, in that only certain cells reproduce while the remainder are in a vegetative condition. By a simple involution of a form such as this, we reach an organism remarkably similar to the planktonic larvae of marine worms, jellyfishes, starfishes, sea anemones, and the Chordata – ancestral to the vertebrates. What regulates the differentiation of cells, originally homogeneous, into different tissues, is still not understood. However, we need not dwell on this in a work on ecology. Let us see how the organisms live and what part they play in the earth's economy.

When we think of the protozoa, we have to realise that we are dealing with living cells which possess basically the same limitations as the cells of metazoa, the cells of our own bodies. The myriad chemical systems used by bacteria and fungi are not available to them. The nutrients they require are in many respects similar to those we need. If we cannot synthesise a certain vitamin, the chances are that protozoa cannot do so either. Their general energy pathways are similar to our own. They are thus not important in ecology in processes of decay. Their main importance is probably in the seas.

On land, the vegetation is mainly rooted in the soil in the form of grasses, trees, shrubs and other plants. In the oceans, this is only possible in limited habitats, where the water is shallow and where light can penetrate in sufficient amounts to be trapped and used. Consequently, by far the greatest part of marine vegetation is in the form of photosynthetic algae and phytoplankton, which are floating in the upper layers of the water – above 300 metres. Of this, there is a tremendous biomass, whose photosynthetic activities exceed those of terrestrial plants. Thus a great part of the earth's atmospheric oxygen is returned to the atmosphere through the agency of these organisms, comprising mostly the diatoms, and the green flagellates.

This vegetation is grazed by other micro-organisms, which are the primary predators of the food chain or web. The process is similar to that on land, although only one-celled organisms are involved, except that the growth is so prolific as to be used for food even by the largest animals on earth, the baleen or rorqual whales.

Fig 10 SIMPLE COLONIAL PROTOZOON WITH LITTLE ORGANISATION. *Conium pectorale:* colony of sixteen individuals, each with two flagella (x c480). (a) surface view, (b) side view, N nuclei, cv contractile vacuole, st stigmata.
From Minchin. 1912. After Stein

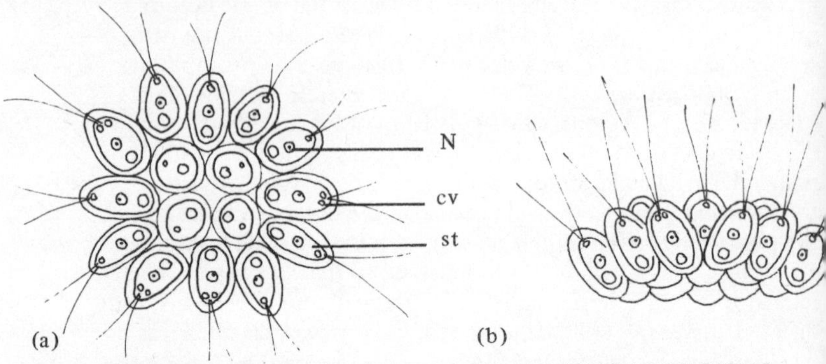

Fig 11 SINGLE-CELLED PROTOZOA FROM WHICH COLONIAL PROTOZOA ARE DERIVED

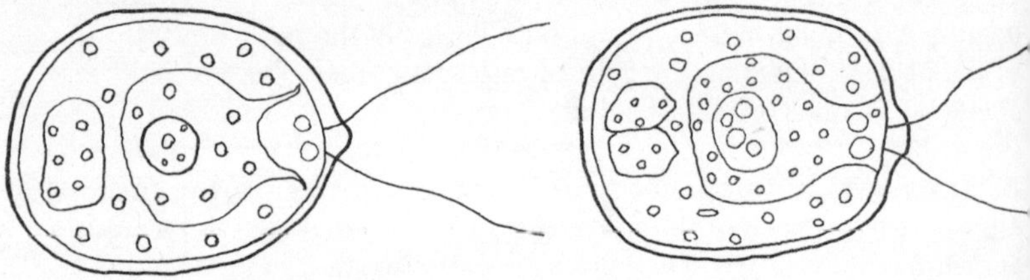

Fig 12 STRUCTURE OF A COLONIAL PROTOZOON. The structure of *Volvox,* showing the different arrangement of the cells in the two species *Volvox globator* and *Volvox aureus.*
After Janet and Klein

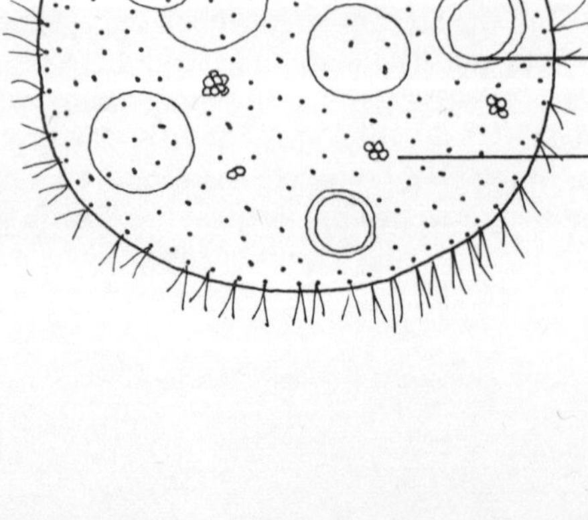

Volvox aureus

Fig 13

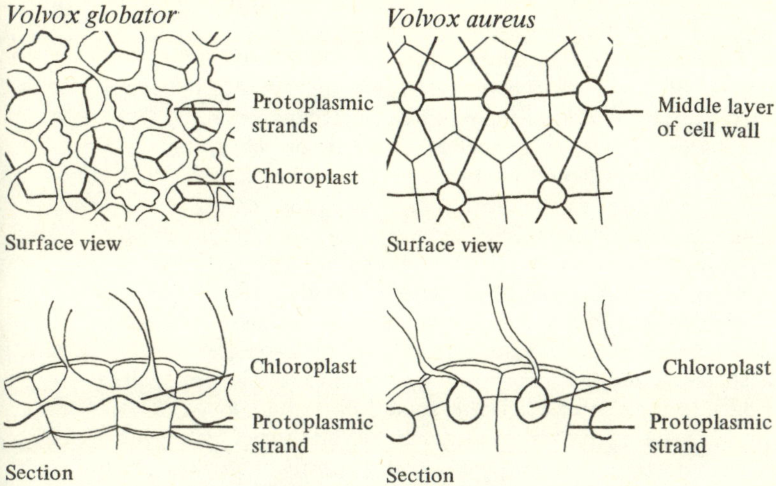

Fig 14 DEVELOPMENT OF METAZOAN LARVAE. Cleavage and gastrulation. There are three types of gastrulation shown here: invagination in *Littorina,* epiboly in *Crepidula,* and ingression in *Patella.*

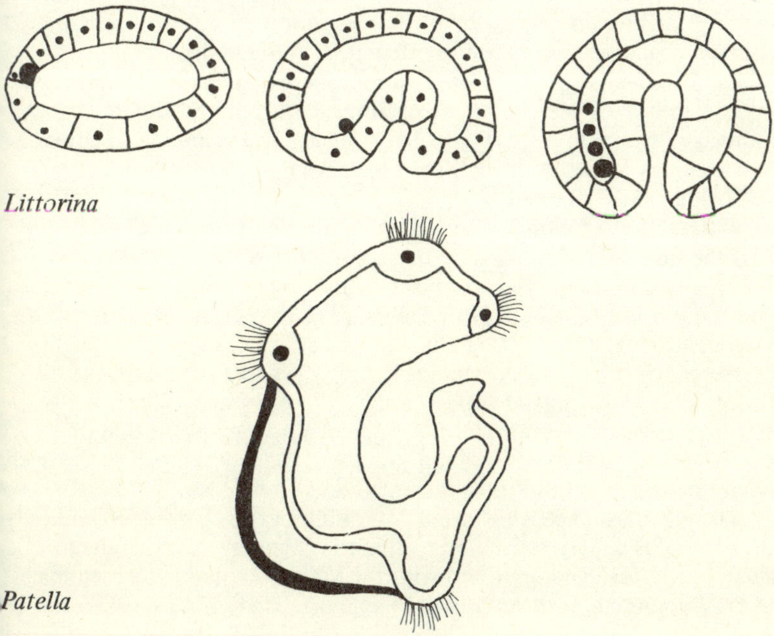

Thus the aquatic food chains or webs, except where waters are shallow, start with these microscopic organisms, which ultimately support the fish and other populations of the seas.

On land, green algae and protozoa are abundant in the scum of ponds, ditches, and stagnant waters, playing a part in the ecology of such habitats. Amongst the predatory protozoa are colourless flagellates, ciliates, and amoebae. These are present in soils in large numbers, where they prey on the bacteria. They in their turn will provide food for earthworms, predaceous fungi, and other soil-living fauna. In terrestrial habitats, nevertheless, protozoa and algae are of comparatively minor importance, except for their activities as symbiotes and parasites of plants and animals.

Every great group of protozoa has produced parasitic forms which are of medical importance. From the flagellates have arisen the leishmanias and the trypanosomes, which cause diseases such as kala-azar and sleeping sickness; some forms of amoebae cause amoebic dysentery; some ciliates are serious bowel pathogens. Finally, there is a whole class, the sporozoa, no members of which have any other way of life than the parasitic. Amongst the sporozoa are the malaria parasites, which are responsible for more human deaths than any other organism of disease.

In summary, the protozoa exploited to the full the opportunities open to the unicellular eukaryote way of life. Their main importance lies in providing the vegetation of the seas, and without them life in the seas would be far less prolific than it is; in doing this, they serve to stabilise the oxygen level of the atmosphere. On land, they have taken largely to a parasitic mode of life and are amongst the most important agents of disease in higher animals. The main line of advance in the eukaryote world lay with the metazoa, for which the eukaryote cell is splendidly adapted.

Following the Cambrian period,[2] colonisation of the seas was rapid: first the invertebrates, followed by the vertebrates as cartilaginous fishes, and later re-invaders from the land or fresh water, reptiles, birds and bony fishes. The heyday of the fishes started in the Silurian and the Devonian, some 450 to 400 million years ago, the earliest forms being cartilaginous fishes like the sharks. During the early Devonian, the first traces of life on land appear in the form of vascular plants and a few invertebrates. Vertebrates first appeared on land at the end of the Devonian, 400 to 350 million years ago; there developed successively amphibians and reptiles, then mammals and birds.

The great age of reptiles lasted 450 million years. Its sudden and drastic end is widely attributed to climatic changes, which rendered them powerless in competition with the better adapted early mammals. However, this supposition is hard to support. There is no satisfactory evidence that a sudden climatic change occurred at the end of the Mesozoic, and mammals at that time were small, insignificant little creatures, which could certainly not have been in competition with the reptiles. Furthermore, this theory does not explain why the aquatic reptiles should have disappeared at the same time as the terrestrial.

The disappearance of the reptiles is one of the great mysteries of earth history. However, it happened some 100 million years ago and the radiation and expansion of the mammals into the vacant ecological niches was in full swing in the Eocene epoch some 50 million years later. They inherited a world that was temperate and warm, but in which trends were towards increasingly cool climates. Climatic deterioration was accelerated during the latter third of the 50 million years of the Cenozoic era, culminating in the glaciations of the Pleistocene epoch 2.5 million years ago. The causes of the great glaciations of the Pleistocene are not known. There were four major glaciations interrupted by warmer interstadials. The last glacial epoch started to recede some 20,000 years ago. The glaciations had a profound effect on the ecology, both as regards plant and animal life. In the tropics, the glacials were represented by pluvials with excessive rainfall, but even at the Equator the glacier limits descended to 5,000 feet. Even at the beginning of the human era, the Sahara was mostly fertile and inhabited by nomadic peoples who hunted and kept cattle. Since then, increasing desiccation has been accelerated by human malpractice. Man himself not only survived the glaciations, but they appear to have suited his way of life as hunter and predator in small bands. Wildlife was abundant, following the ice fringes north in Summer and south in Winter; meat supplies were easily won by hunting or trapping. During the winter, the meat could be deep frozen in ice or snow, so there were no shortages. Animal furs and skins provided warm clothing and the marvellous invention of fire kept the caves or mammoth-tusk hutments warm and snug.

Man's way of life was overthrown when the climate became warmer in the Mesolithic era, for the forests spread over the land. Hunting became more difficult and less rewarding and changes of diet were necessary for survival. Then came the agricultural revolution with the first hesitant attempts to till the land and grow grain and domesticate stock.

The profound effects of climatic change, then, were something that had not been experienced since the early Cambrian nearly 550 million years previously. It is now 10,000 years since the last glaciation receded. Are we enjoying an interstadial after which the ice will return? Or are we entering another prolonged period of temperate climate? If the ice is to return, when will this happen? To these questions, we know no answer. A return of the ice to its previous southern limits along the line of the Thames and south of the Alps would render life as we know it impossible, and the weight of the ice on land surfaces would tilt them and cause widespread flooding.

Meanwhile, it behoves us to study and overcome our problems as we find them, but with the thought that nature is more powerful than we are and will inevitably have the last word.

NOTES

1 See J.E. Smith, 'The Nutrition of Bacteria and Micro-Fungi', in R.N.T -W-Fiennes, *Biology of Nutrition,* Pergamon Press, Oxford, 1972, and S.H. Hutner, H. Baker, O. Frank and D. Cox in Fiennes, op. cit.
2 See E.C. Olsson, 'Climatic Change and its Influence on the Evolutionary Development of Life Forms', in Fiennes, op. cit.

Further Reading

Borradaile, L.A., and Potts, F.A., *The Invertebrata,* University Press, Cambridge, 1958
Clark, R.B., *Dynamics in Metazoan Evolution,* Clarendon Press, Oxford, 1964
Cloudsley-Thompson, J., *The Temperature and Water Relations of Reptiles,* Merrow Publishing Co., London, 1972
Mercer, E.H., *Cells and Cell Structure,* Hutchinson Educational, London, 1961

4 PHOTOSYNTHESIS AND THE ENERGY PATHS OF LIFE

In the preceding pages, continual reference has been made to energy and its modes of use in living things. It will be well now to pause for a consideration of the mechanics of life and its power sources. The system is often spoken of as 'energy flow'. Energy is trapped into the system and made to flow upwards into energy stores, from which it is released at need to provide the driving force of metabolic processes. The energy can be studied in relation to a single plant or animal, or groups of plants or animals; that is, energy flows can be traced through a habitat and its 'associes', through the primary producers, predators and scavengers back to the products of decay and recycling of the elements.[1]

In the preceding pages also, reference has been made to a property of living systems, by which increasing order emerges from disorder. Habitats under the dictates of ecological laws become ordered; living systems were initially varied, but have become stereotyped; the probiotic earth was disordered but has acquired increasing order; the atmosphere has developed a constant composition. The Second Law of Thermodynamics decrees that natural processes will flow towards an increase of 'entropy' or randomness until equilibrium is reached; flows are downhill, not uphill. A hot body will yield heat to a cold, until the two reach the same temperature. A high pressure system will yield pressure to a low pressure system. These principles regulate equilibrium in meteorological conditions and between oceans and atmosphere. In biological systems, however, energy flows move towards greater and more complex organisation against the dictates of the Second Law. This can only occur because of the acquisition and use of energy acquired from natural sources, the chief of these being the radiant energy of the sun.

We have seen how the shortwave radiation of the sun in probiotic days could have, and no doubt did, mediate the synthesis of 'biochemicals' of high molecular weight. The most primitive of living organisms are those which can only obtain energy by degrading biochemicals – in some case chemicals – thereby releasing and using the energy that had been stored in them. This process would be self-limiting, because the ultra-violet light by which they were synthesised would become less available as oxygen was released and shaded the earth's surface; the ultra-violet light could not have penetrated more than a few metres through water. The development of chlorophyll by living systems led to two important changes: first,

Fig 15 THE FATE OF SOLAR RADIATION. Energy intake at the earth's surface at midday.

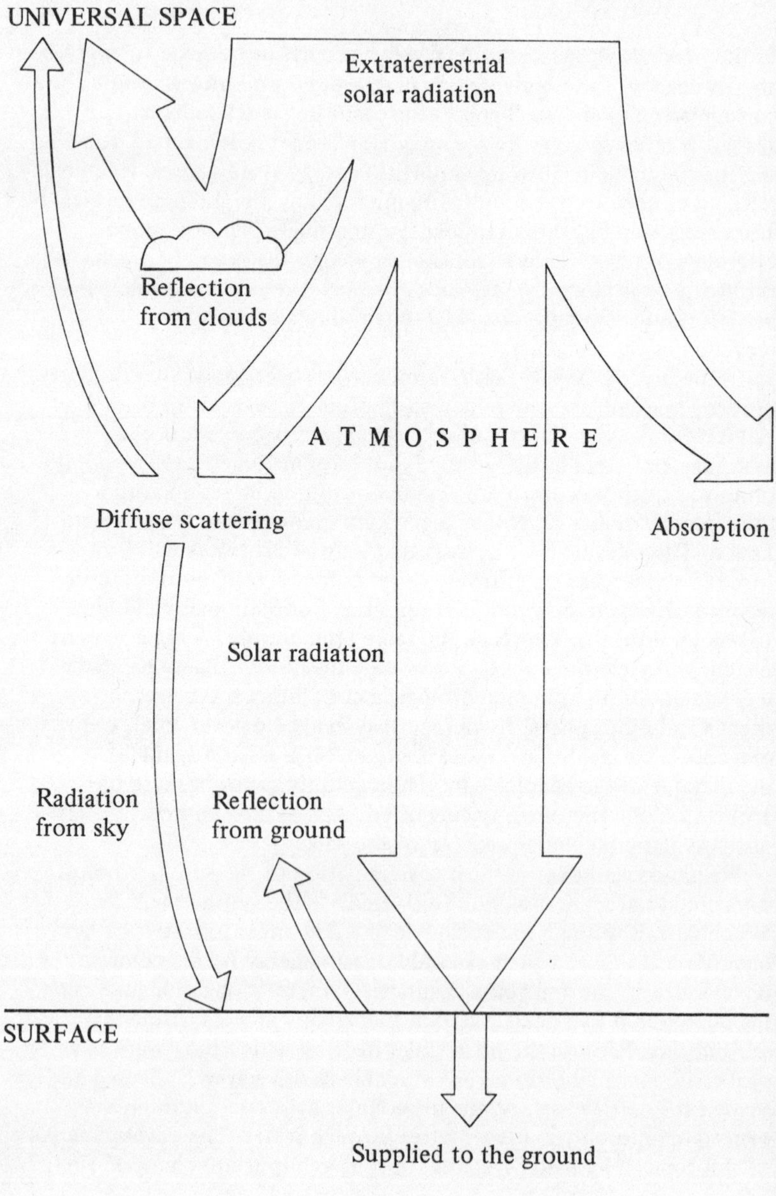

use was made of the visible light rays as the prime energy source; secondly, the trapped energy (electrons) was made to flow into energy stores, from which it could be recovered and converted from chemical energy back into work energy. In biological systems, energy occurs in four forms: (1) chemical or stored energy; (2) electric energy as in nerve transmissions; (3) muscular energy; (4) osmotic energy, as in glandular secretions. A biological system, therefore, must be able not only to acquire energy, but also to direct it at need to any of these four forms. While the basic principle of photosynthesis and energy flow is simple enough, the mechanics are somewhat complex.

So when the earth's surface became screened from the shortwave ultra-violet light of the sun, life would not continue unless an alternative power source were developed; for this, those wavelengths which still reached the surface must be used. The only suitable radiation for the purpose was that of the visible spectrum, and this involved many complexities. However, a system capable of trapping light quanta (photons) was developed, first in the green bacteria, and then further developed in the chlorophyllous protozoa and in the green plants. This development led to the increase in atmospheric oxygen, which enabled the evolution of metabolic systems based on oxygen, and so to the appearance of higher life forms as we know them.

Photosynthesis

Matter, as all know, is composed of atoms, and atoms are composed of a 'nucleus' carrying a positive electric charge round which whirl one or more electrons having a negative electric charge. The path of the electron round its nucleus is known as its 'orbital'; the lowest orbital as the *a* orbital, and higher orbitals are *b, c, d* and so on. The simplest atom is that of hydrogen, of which the nucleus consists of a single 'proton', around which orbits a single electron in the *a* position. The electron is prevented from merging with the nucleus by its own kinetic energy, that is the energy of its own motion; it cannot fly away from the nucleus, because of electrostatic forces. However, if an electric field is applied to the atom, the electron can be raised to a higher orbital, theoretically to an infinite distance. This requires an input of energy, and this energy can be recovered if the electron is again dropped back to its lowest orbital. The energy required to remove an electron to infinity is known as the 'ionisation potential' and it can be measured in electron volts: 1 eV/atom = 23.06 k cal/gram atom* which gives us a relationship between electric force and heat. Electrons at higher orbitals require less energy to remove

*In more modern literature, usually converted to S1 units, which would be without meaning to non-scientists – and indeed to most biologists.

them than those at lower orbitals, a point that is important in photosynthesis. The orbitals of electrons are less simple than here stated, but the Bohr model, as outlined, will suffice for our purposes.

To use visible light as a power source requires the sun's photons to be trapped and their energy used to raise electrons to higher orbitals thus creating a field of electric energy. However, since electronic energy is unstable and cannot be conveniently stored, the further problem arose to convert the unstable electric energy into stable chemical energy. The third problem was to reconvert the stored chemical energy into electric, when it was needed to power life processes, that is to steal the light quanta and manipulate them through biological systems. So, when we study the use of energy by living systems, we are in fact tracing the travels of these light quanta not only through an individual cell or system of cells, but also through an entire ecological system, plant → herbivore → predator → secondary predator → systems of decay and disintegration.

When light rays of the red end of the spectrum fall on the green pigment chlorophyll *a,* two electrons in outer orbitals of its magnesium atom are raised to higher orbitals. Unless they were trapped in some way, the electrons would fall back to their original positions in a minute fraction of a second giving a flash of fluorescence. It is as if a heavy ball were shot upwards by means of a spring. Normally, the ball would quickly fall back to its original position. However, a device could be arranged whereby the ball would be trapped at the summit and then made to fall back through a channel having a series of steps. At each step, a lever or switch could be arranged to be activated during its descent, and in this way the energy of its fall, originally derived from the spring, could be converted into the forms of energy important in biological power systems, mechanical, electric, or chemical (osmotic). In photosynthesis, the excited electrons from the chlorophyll's magnesium are diverted to a high energy chemical acceptor system, the substance NADP being converted by addition of an additional hydrogen atom to NADPH. The source of this hydrogen is important, as we shall see.

At this stage, two electrons need to be replaced in the magnesium atom, and a hydrogen atom must be found for the NADPH. In more advanced photosynthetic systems, this is achieved by splitting the water molecule and temporarily detaching electrons from the hydrogen atom. However, let us look first at the more simple system as in the green bacteria. In these, the NADP → NADPH acceptor system is unnecessary. Electrons are detached from sulphur, a ready electron donor; they are transferred by a chemical transport system to the chlorophyll to replace the electrons lost by the magnesium through the action of red light; those electrons then fall back to the sulphur, some of their energy being trapped to convert ADP → ATP. The basic material used by the green bacteria is H_2S (hydrogen sulphide), so that the end point is molecular sulphur. Certain enzyme systems are also involved. The chemical reaction goes like this: $CO_2 + H_2S \rightarrow 2CH_2O + 2S$. The CH_2O is the basic carbohydrate unit

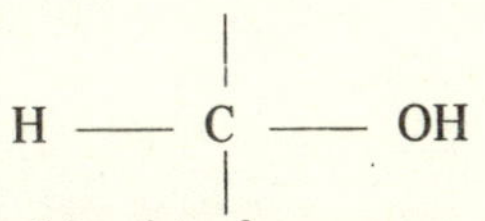

and these units become combined to form sugars. In the end reaction, the energy trapped in the ATP is used to reduce CO_2.

The use of H_2O as the source of electrons and hydrogen presents greater problems, since the red waves of the spectrum lack sufficient energy to split the water molecule. This is done by the action of blue light trapped by chlorophyll *b* in a second centre. Thus, two photosystems are used, the original one photosystem I, using red light and photosystem II, using blue. By this means, the water molecule is split and free electrons derived from hydrogen are fed into the system to be transported by a chain of chemical electron donors and acceptors, until they reach the chlorophyll *a* in photosystem I. The result is the liberation of free oxygen into the atmosphere. The chemical reaction goes like this: $CO_2 + H_2O \rightarrow CH_2O + O_2$. Expressed another way: $6CO_2 + 12\ H_2O$ + 673 calories + chlorophyll would produce $C_6H_{12}O_6$ (6 carbon sugar) + $6O_2 + 6H_2O$. Meanwhile, the hydrogen protons pick up the electrons from the chlorophyll magnesium and enter into the reaction NADP → NADPH. NADPH in its turn reduces CO_2 to carbohydrate.

The only actual input of energy into the system is that derived from the red light in its action on chlorophyll *a*, although blue light plays a part. No pigment, other than chlorophyll *a,* reacts to light rays in this way; therefore, all life dependent on oxygen depends on this one pigment. In actual fact, little light of the visible spectrum is wasted by the chloroplast system. The chloroplast is a kind of cell within a cell, and contains numerous 'grana', small organelles, which are bathed in cell sap. The grana are composed of numerous microscopic leaflets with a large surface area exposed to the cell sap. They are packed with pigments of many colours, ('the autumn tints') as well as the green chlorophylls, and by this means absorb most of the incident light. The light rays set up a state of 'resonance', and by the system of 'resonance transfer' their energy too is transferred to chlorophyll *a* in a form which can be utilised in photosynthesis. Thus all these pigments, chlorophyll *b* included, act as 'antennae' for the chlorophyll *a.* The grana are also rich in the accessory enzymes and compounds necessary for electron transport and other reactions, and

Fig 16 ENERGY PATHWAYS IN PHOTOSYNTHESIS

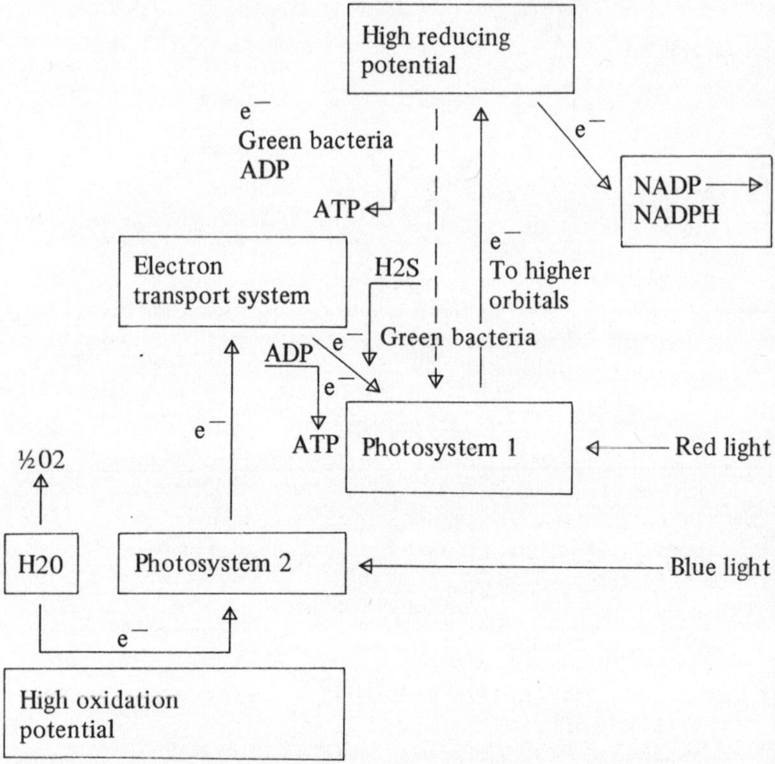

Fig 17 THE GRANA ARE COMPOSED OF MANY FINE LEAVES WITH A LARGE SURFACE IN CONTACT WITH THE CELL SAP. Fine structure of a chloroplast. The organelle is enclosed by a single membrane and contains numerous clusters of parallel membranes called grana which support the photosynthetic mechanism.

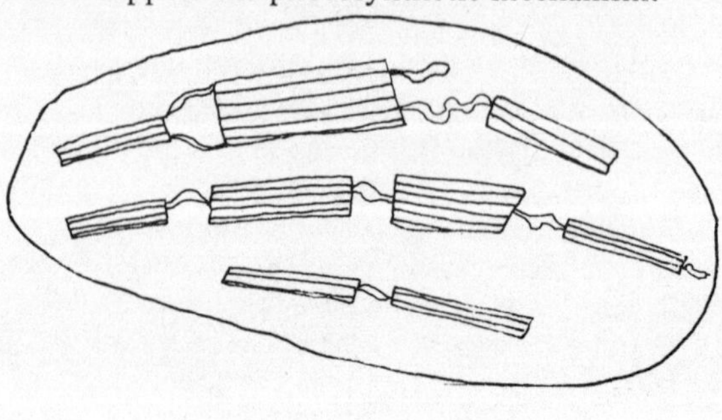

provide ready exchange of materials with the cell sap so that a rapid turnover is assured.

Nevertheless, the efficiency of light harvesting is limited by two factors, if not more. Photosynthesis is, of its nature, biphasic, requiring first the synthetic phase, and secondly a recovery phase while the synthesised products are exchanged for raw materials. Experimentally, its efficiency can be greatly improved in terms of rapidity of plant growth if the plant is exposed to rapidly intermittent light, allowing for optimum periods of synthesis and recovery. In the second place, photosynthesis is most efficient at certain optima of incident light and temperature. One might suppose that the high intensity light of desert areas would assist rapid photosynthesis. In fact, this is not the case, and many desert plants have silvery layers in their leaves which protect the chloroplasts from the full light intensity. Because of these two factors, theoretical calculations of the productivity of incident solar light on the earth's surface are of little value.

The end product of photosynthesis can be proteins or fats in addition to carbohydrates.[2] The process of photosynthesis is thus a matter of some complexity. In fact, the conversion of CO_2 into higher molecular compounds is not unique, since this can occur in the liver as part of normal metabolic processes powered by ATP. It is the use of the light photon as the energy source which makes the process unique. The whole of life as we know it today is dependent, therefore, on one green pigment. The chlorophylls themselves, however, belong to a well-known group of biological pigments, the porphyrins. The red pigments of the blood, responsible for the exchange of CO_2 and O_2 in the lungs and tissues, also belong to this group, but contain iron instead of magnesium. Primitive metazoan animals such as snails and tunicates experimented with other oxygen-transporting porphyrin pigments containing copper, niobium and vanadium.

From the high energy levels in carbohydrates and fats attained as a result of photosynthesis, the electrons when used in living systems drop back via ATP in a controlled and organised way stepwise to their lowest orbital positions; these positions occur in water, which is thus the excretory end product of energy in living systems. At each step energy is made available for life processes. Electronic energy could not, however, by itself be the basis of higher living systems, because it cannot conveniently be stored. It must be converted into chemical energy, and this is what happens when the energy is transferred to NADP and stored in the phosphate moiety.

Energy is, as we have seen, transferred at the end point of the photosynthetic process into carbohydrates and fat, which being insoluble can be accumulated. It is transferred back by way of ATP, and the reverse of photosynthesis occurs, 'oxidative phosphorylation'.

Electronic energy is the high biopotential of free electrons; chemical energy is the collective energy of the electrons forming the molecule. The single electron is mobile and can go to other atoms or molecules, that is to other orbitals of lower energy and with its negative electric charge will create an electric field. Chemical energy creates no such field and the electrons are immobile, being linked to molecules.

Energy Flows in Ecological Systems

The most characteristic features of life are these energy transformations, in which chemical energy is transformed into mechanical, electric, or osmotic work as in muscle, nerve, and glands respectively. How this is done is by no means simple and will not be discussed in detail here. Basically, the system depends on the balanced flow of electrons between systems of donors and acceptors mediated by enzymes. Donors possess loosely linked electrons; acceptors have spare capacity. The living cell is essentially an electrical device; in it the macromolecular structure is the framework in which the transduction of electrical energy into mechanical work takes place. The sources of electrons are the *activated* electrons of metabolism and are donated from the paired electrons of nitrogen, oxygen, and sulphur; the main acceptor group is found amongst carbon compounds.

The transmission of electrons from one substance to the other means the oxidation of one (the down path) and the reduction of the other (the up path). If solutions of reductant and oxidant are placed in beakers, the electrons can be made to pass from one to the other through a wire. The potential difference between the two solutions can be measured and gives information about the free energy exchange of the oxido-reduction which would take place if the solutions were mixed. The 'redox potential' allows us to calculate the energy changes taking place in the oxidation cycle. The redox does not measure the speed of a reaction, only the amount, and the energy change may differ with the medium. Mitochondria responsible for initiating and regulating these changes in cells, the medium is probably lipid, complex fatty compounds..

When energy needs to be recovered from the energy stores, the reverse of photosynthesis takes place, known as oxidative phosphorylation. In this oxygen is necessary, and the stored materials are again degraded to CO_2 and H_2O. For instance the group $-CH_2-$ present in fats and other biochemicals plus $3O \rightarrow CO_2 + H_2O$; the CO_2, whether from plants or animals, is returned to the atmosphere. This is a similar process to that which occurs when coal or wood is burnt, but in combustion processes the stored energy is dissipated as heat, whereas in biological processes it must be diverted to processes of metabolism. In warm-blooded animals (homoiotherms), some

Fig 18 ATP could act as the 'transformer' of electric energy to chemical energy. The electrons would be trapped by the adenine and passed to the phosphate to be incorporated into the high energy phosphate bonds. (simplified formula)

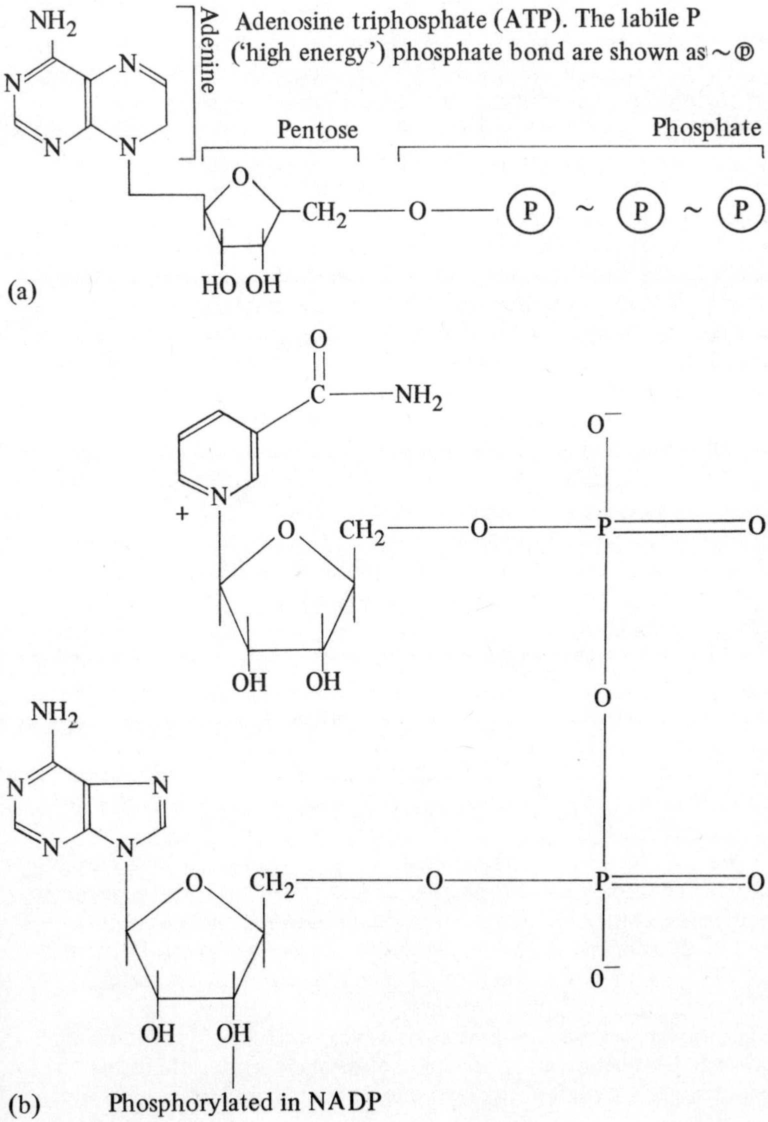

energy is required for maintenance of body temperature and some of the heat produced is dissipated in maintaining this at constant level. This is why warm-blooded animals require a constant intake of food, whereas the larger cold-blooded animals (poikilotherms), such as crocodiles or pythons, feed only once a week or so and can survive for months on end without any food at all.

The use of energy for metabolic purposes depends on the extraordinary substance ATP, of which mention has already been made. ATP is a complex molecule, so arranged that it can act as a 'transformer' of chemical into electrical energy or vice versa. When fats or carbohydrates are broken down by enzyme action, the falling electrons, instead of creating heat, convert ADP (adenosine disphosphate) into ATP (adenosine triphosphate) in which the energy is trapped as chemical energy in the phosphate; it can then be delivered back as electrical energy, when and where it is required. ATP is unsuitable as the main energy store, but acts admirably as the medium of exchange. Biological 'currency' is stored in the high denomination notes, represented by fats and carbohydrates; ATP represents the small change of current transactions. ATP (adenosine triphosphate) consists of a compound base adenine, linked to a phosphate chain by a pentose (5 carbon sugar). This linkage is so arranged that various foldings are possible. Adenine is a good electron donor and the P atom, with unoccupied *d*-orbitals, would be a good acceptor. Thus electron exchange between the two could occur, if the molecule were folded in the right way. There is good evidence that the molecule is folded in such a way that the transfer could take place, and this explains the transformation of the electrical energy taken on by the adenine being converted to chemical energy in the phosphate.

It has been worth while to give this simple account of the trapping and use of radiant energy in biological systems, both because of its inherent interest as our basic life force and because it has been necessary throughout this work to make reference to it. For a more detailed account, the reader cannot do better than to study various contributions to the book *Biology of Nutrition,*[3] especially that by T.E. Cartwright.[4]

We have seen hitherto how light energy is used to convert simple chemical compounds, carbon dioxide and water, into storage elements, and how it can be recovered for building the structural elements of life, and made available for use in metabolic processes. Provided provision for decay is present in the form of lower saprophytic life forms, life could go on, and the elements could be recycled without higher life forms, apart from plant life. However, ecological systems are not constructed in this way, since there exist many higher life forms which use plant material as their energy source, and other life forms which live on the plant feeders. Mammals themselves are singularly ill-adapted to live on plant material. They were evolved in the Eocene and Oligocene epochs some 50 million years ago from primitive insect-eating ancestors. They do not have the digestive apparatus to deal with plant material, and lack the enyzmes necessary to split the higher polysaccharides such as cellulose, of which plants make their cell walls and largely consist. However, mammals have produced their herbivorous forms in the ruminants and the horse family, amongst the rodents, marsupials and even man. None of these could use plant material unless assisted by lower organisms which have not lost their power to attack plant material.

It is here of interest to see what happens to the energy of the photons trapped by the plants when introduced to a comparatively simple digestive system, such as that of a carnivore, which does not depend on lower life forms to do the job for it. The point is important in relation to the study of energy flows through ecological systems, which follows.

Before becoming available to build animal tissues, plant material must again be degraded by herbivores to comparatively simple compounds which can be absorbed through the bowel wall into the body. It is then resynthesised in the body, reconverted into storage and structural materials, and the energy is again released for metabolic activities. The digestive system of the flesh-eating animal again splits the proteins into their component amino acids; twelve carbon sugars, such as sucrose (cane sugar), are broken down to six carbon sugars; and fats are broken down to fatty acids and glycerol. Once absorbed, these materials again enter the metabolic pathways to be degraded, with release of their energy, to the metabolic end products of carbon dioxide and water.

Both the digestive and the anabolic (rebuilding) processes require considerable inputs of energy, so that at each stage a great deal of energy is lost. Furthermore, the whole process is subject to the First and Second Laws of Thermodynamics. A given quantity of organic matter contributes precisely the same amount of energy, whatever are the means used to degrade it; and when put to work an identical amount of heat will be produced whatever the nature of the work. No work can be performed without the conversion of some energy from a relatively non-random form (chemical or potential energy) to relatively random energy (heat) which is lost. We are, thus, going to find energy loss at each step in our ecological food chain, plant to herbivore, herbivore to carnivore, carnivore to scavenger, and from all these to organisms of decay. A further considerable loss of energy occurs also in animal excreta, since digestive processes never extract all the energy-containing materials available. This is of lesser consequence when excreta are

Fig 19 BIOMASS AND SOLAR ENERGY

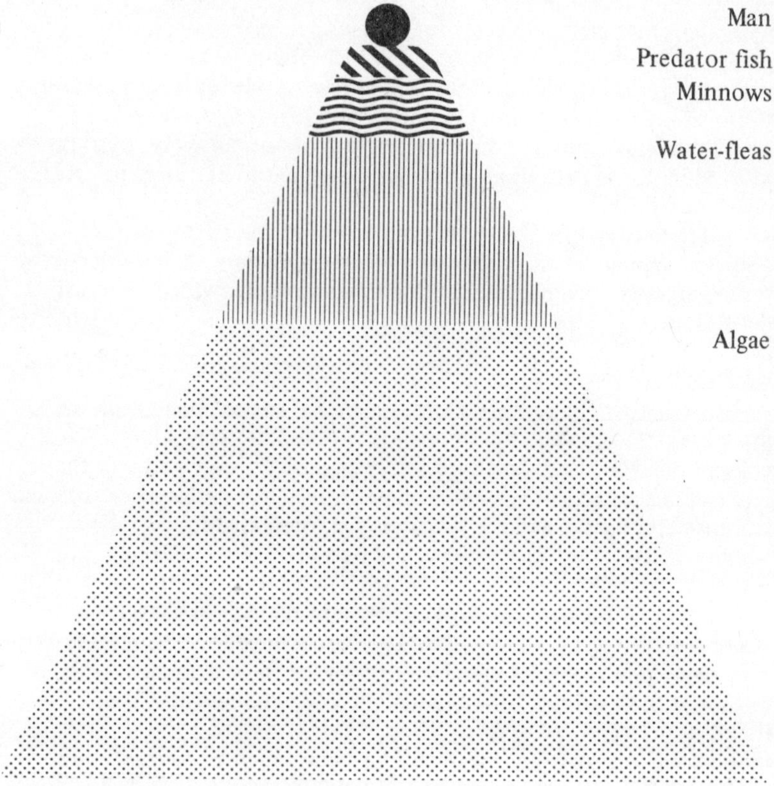

returned to the land, but becomes more important when they are swept out to sea in the main drain. Obviously, when population pressures on a habitat are high, the most efficient management of the habitat becomes important.

Just how much energy is lost in the process is seen from the accompanying diagram (Fig. 19) from Hicks.[5] This shows at least that theoretical possibilities exist of more efficient use of energy resources, but at each stage in the food chain the more valuable elements become more concentrated, so that the supposed loss of efficiency is in part at any rate illusory, a point that is often overlooked. Animal products are expensive and the poorer human communities have mostly adopted diets preponderantly vegetarian; nevertheless they are a more valuable food source.

A natural ecosystem may be more or less productive, in the sense of the total biomass it will produce. Its total production will depend on the efficiency of its individual members. Basically, however, this will be a function of the climate and the soil, and its productivity can be improved by improving one or both of these, by supplying water or protecting surfaces from evaporation, or by providing missing nutrients to the soil. It could also be improved by replacing inefficient producers in the ecosystem with more efficient ones. For instance, there is a tendency today to suppose that the introduction of cattle, as the main source of animal protein, to tropical Africa was a mistake made by early immigrant peoples. It is argued that other indigenous herbivores, such as eland and buffalo, are natural and more economical exploiters of the ground cover and would cause less damage to it. Schemes have been suggested for replacing cattle with these animals, though I do not know of any studies made on the energy flows that occur when this is done. Nevertheless, it is certain that man has it in his power, by making proper studies of habitats and ecosystems, to improve energy flows and thereby gross total production. Agriculture could, therefore, lead to greater efficiency rather than less. The same remarks apply with equal or greater force to aquatic environments, especially estuaries. This has been shown, particularly by the Japanese in their fish culture ventures.

However, it is worthwhile looking at a few figures, drawn from Lamont C. Cole.[6] The supply of solar energy to the earth amounts to 13×10^{23} gramcalories per year, equivalent to a continuous power supply of 2.5 billion billion (American billion) horsepower. One third of this energy is immediately reflected back to space, chiefly by clouds. The rest is on loan to earth, until it is radiated back again. It serves to melt ice, to warm the land and oceans, to evaporate water, to generate winds and waves and currents. Only about four-hundredths of one per cent of the solar energy is used for photosynthesis. Of this, the plants use one-sixth for their own metabolism, leaving five-sixths available for animals and other consumers. Even of this, some 5 per cent is dissipated by forest and grass fires and burning of plant products for fuels. Of the plant material, herbivores succeed in converting only 50 per cent of the calories, and of this only 20 to 30 per cent is built into the protoplasm giving an efficiency of 10 to 15 per cent. The conversion efficiency of carnivores is around 20 per cent, because there are fewer indigestible residues.

Cole gives an interesting illustration taken from Lake Cayuga. Of every 1,000 calories stored by the lake algae, only 150 calories can be converted into protoplasm by small aquatic animals. Smelt, which feed on these animals, produce only 30 calories. If the smelt is eaten by a trout, the figure is reduced to around 6 calories. When the trout is eaten by a man, only 1.2 calories are left. Cole goes on to estimate that, if the existing population of man were to feed exclusively on vegetable material, he would consume almost exactly 1 per cent of the total food resources of the earth. Since there are more than one million species of animals, he feels that this is an unreasonably high proportion. If man continues to demand a meat diet, this, of course, represents a much higher proportion of available resources.

One would like to see an analysis of energy flows through a highly industrialised society, such as that of the British Isles. In such a society, special features operate which may in the long run lack stability. Such are the importation of food and fertiliser from overseas countries. In addition, much residual energy and other valuable resources are wantonly swept to the sea in the sewage system. The latter is discussed by J.C. Wylie in his admirable little book *Fertility from Town Wastes.*[7]

Wylie, the Divisional Engineer of a Scottish burgh, was the architect of a sewage composting system. He succeeded in persuading his authority to give a trial to a disposal plant in which the garbage was composted with sewage sludge. The compost was sterilised by fermentation heat and was then sold for use on farms at good prices. Further extraction retrieved valuable minerals, chiefly phosphate and nitrogen, which are responsible for estuary pollution.

By recovery of waste materials in this way, industrialised countries could do much to redress an adverse balance in their energy accounts, and indeed in their balances of valuable minerals. They could also, with profit to themselves, avoid polluting their inland and inshore waters. While Wylie's initiative has been followed by some enlightened local authorities, argument still continues amongst local councillors as to whether a penny rate (the system in which the UK funds local schemes) should be spent on the necessary plant, a penny rate recoverable by the sale of the compost, or whether to spend a twopenny rate on shipping the garbage in barges to destroy the marine ecosphere.

I have touched briefly in this rather discursive chapter on natural minerals required in ecosystems. Further discussion of these and of the natural cycles by which they are reused will be left to a later chapter. Our main concern here has been to see how the sun's energy is trapped to become the driving force of life and how it flows out not only through living systems themselves but also through ecological systems. If man arrogates to himself the right to change the earth's ecosystems, he should at least acquaint himself with the best ways of controlling this energy flow for amelioration of his habitat rather than to its detriment.

NOTES

1 I.N. Healey, 'The Habitat, the Community, and the Niche', in R.N.T -W - Fiennes, *Biology of Nutrition,* Pergamon Press, Oxford, 1972

2 H. Davson, *A Textbook of General Physiology,* Churchill, London, 1959, Ch. XXII, 'Photosynthesis'
3 Fiennes, op.cit.
4 T.E. Cartwright, 'The Ways of Acquiring and Utilizing Energy', in Fiennes, op.cit.
5 Sir C. Stanton Hicks, 'The Nutritional Requirements of Living Things', in Fiennes, op.cit.
6 Lamont C. Cole, 'The Ecosphere', in R. Ehrlich, J.P. Holden, and R.W. Holm, *Man and the Ecosphere,* W.H. Freeman & Co., San Francisco, 1971
7 J.C. Wylie, *Fertility from Town Wastes,* Faber, London, 1955

Further Reading

R. Ehrlich, J.P. Holden, and R.W. Holm (eds), *Man and the Ecosphere,* W.H. Freeman & Co., San Francisco, 1971
E. Le Cren, and M.W. Holdgate (eds), *The Exploitation of Natural Animal Populations,* British Ecological Society Symposium, Blackwell, Oxford, 1962
A. Szent-Gyorgyi, *Bioenergetics,* Academic Press, N.Y., 1957
A. Szent-Gyorgyi, *Introduction to Submolecular Biology,* Academic Press, N.Y., 1960
A. Szent-Gyorgyi, *Bioelectronics,* Academic Press, N.Y., 1968

5 THE OCEAN/ATMOSPHERE SYSTEM AND 'CYCLING' OF THE ELEMENTS

Ecosystems may be wasteful and yet the energy flow through them enables increasing order to be produced from chaos. No less important is the flow of nutrients through these systems; they are the raw materials of which both individuals and the systems as a whole are constructed, and comprise the fuel the systems use. The ocean/atmosphere system, another flow system, recycles water resources. All life is based on water and cannot exist, except in a resting phase, in its absence.

As with life itself, the ocean/atmosphere system[1] requires a power source and, as with life, the power source is the sun. Whereas living systems use the visible rays of the sun, the ocean/atmosphere system uses the heat rays. Basically, water is evaporated from water masses, chiefly the ocean; the water vapour passes into the atmosphere; it is cooled in the upper atmosphere, forms clouds, and is reprecipitated as rain. When rain falls on land surfaces, some of the water finds its way directly into rivers and is returned to the sea. Some of the water is stored in lakes and reservoirs, in the polar ice, in the water table, or in the soil. The stored water may remain in the ground for long periods or be returned to the atmosphere by evaporation or transpiration from plants. On the face of it, this is a simple enough story, but the science of meteorology is by no means simple because air and water movements are influenced by so many unpredictable and unknown factors. In no two years are weather patterns the same, and even in waterless deserts occasional rainstorms occur. On a short-term basis, there are said to be eleven-year climatic cycles related to sunspot activity. On the medium term, there appear to be warmer and colder trends to the order of 400 to 500 years. On the long term, there appears to by no rhyme or rhythm which orders the appearance and disappearance of ice ages.

Since I am not a meteorologist, I shall not attempt to explain that which is obscure even to the experts. However, we may briefly look at the forces at work and review past and present climatic trends. Important forces are: (1) the heat of the sun and its position during the different seasons; (2) the spin of the earth which determines a predominantly westerly system of winds; (3) the gravitational effect of the moon on the earth causing tides; (4) the nature and direction of ocean currents; (5) the effect of land topography on air movements.

The sun's heat, being most intense at the ecliptic, causes moisture-laden warm air to rise; when it is cooled in the upper atmosphere, precipitation occurs. Thus a rain belt follows some weeks behind the sun's position. This belt is known as the inter-tropical convergence zone, and air movements occur from all directions into the zone which is thus cyclonic and constitutes the area of uncertain winds known by sailors as the doldrums. During the summer, much of the air which has passed to the upper atmosphere is transported to the polar regions, whence it is returned in very cold condition in the latter part of the winter as blizzards. At the same time, the westerly prevailing winds distribute the ocean moisture to the land masses. Over the great oceans, the maximum effect of the sun in causing an uprush of moisture-laden air may be so great as to cause hurricanes or typhoons.

Over uniform surfaces, air movements could be reasonably simple, but neither marine nor terrestrial surfaces are uniform. The oceans consist of great masses of water, moving this way or that in warm or cold currents, layered one above the other, downwelling and upwelling, diverted hither and thither by land masses and the shape of the ocean floor. To a great extent they are caused by the melting of Arctic and Antarctic ice by the sun's rays in the northern and southern summers. A cold current releases less water vapour into the atmosphere, and so is conducive to dry conditions. Cold currents of Antarctic origin pass along the west coasts of Africa and South America and result in desert or semi-desert conditions there. Conversely, warm currents cause moist conditions and high precipitation occurs over lands in their vicinity. This accounts for the moist conditions in countries such as Britain and Ireland whose shores are washed by the Gulf Stream.

Any factor which causes moisture-laden air to rise and be cooled, will also give rise to precipitation. Such factors are hills and mountains, and greater precipitation occurs, therefore, in hilly places. This effect is enhanced by the moist atmosphere in such places, the result of transpiration from the generous vegetational cover arising from the high rainfall. The vegetation also prevents excessive runoff and ensures absorption of the water into the soil, from which it seeps into the water table and feeds springs and wells. One of the surest ways of destroying a habitat is to cut down watershed forests, an activity to which man is much addicted. Precipitation is reduced; topsoil is removed by water erosion; and lower-lying land is destroyed by periodical severe flooding. At the same time, rivers come to flow only seasonally, and the water table becomes lowered.

Another factor is often adduced as a cause of climatic change, namely the composition of the atmosphere in respect of carbon dioxide. This problem is discussed at length by G.H. Plass.[2] An increase in atmospheric carbon dioxide is said to have a greenhouse effect tending to make the atmosphere warm and steamy, because the warm air is prevented from rising. It has been suggested that the burning of fossil fuels since 1860 has raised the atmospheric carbon dioxide from

0.028 per cent to 0.033 per cent with a consequent rise of the average temperature in the northern hemisphere by 1 to 2°F. If this were so, one might expect the opposite effect to have occurred in the Carboniferous, when the coal measures were laid down. This does not appear to have happened, and any effect such as this must surely be rather marginal. However, one factor of the ocean/atmosphere system that must not be overlooked is that of the seas in absorbing and regulating the levels of atmospheric carbon dioxide so as to maintain a stable equilibrium; reference has already been made to this (p.21).

Glaciation

The causes of the great glacial epochs, with their interstadials, are unknown. They are usually attributed to fluctuating intensity of solar radiation. There is no proof of this, but it is perhaps the most reasonable explanation. However, the effects of glaciation on land topography are far reaching. At the height of the last glaciation, the southern limits of the ice reached the Thames and to the south of the Alps. Below the permanent ice, the vegetation would be that of Arctic tundra. During the summer, the ice would melt and expose enormous areas of low Alpine-type vegetation, very well watered by streams from the melting ice and snow. The vegetation supported enormous herds of grazing animals, wild horses, mammoths, reindeer and so on, which moved south in winter and north in summer. Man himself was very much at home in this habitat and lived by hunting these animals, at which he was very successful.

The glaciation was accompanied in tropical areas by periods of heavy rainfall, known as pluvials. The equatorial forests extended far to the north of their present limits, and the Sahara was a well-watered plain, over which animals grazed and man also hunted. The ice only began to recede some 20,000 years ago, after which forests grew in man's former hunting areas, the wild life changed in character, and conditions became much less to the liking of our ancestors. It was this which forced them to take to agriculture and stock husbandry, and started the great social revolution which has been going on ever since.

During the glaciations, there was a drastic change in the ocean/atmosphere cycle. Myriads of tons of water were locked up in the polar ice, to an extent that much land was exposed where now there is sea. The English Channel was non-existent, England being joined to France, and much of the North Sea was dry land. Alaska was joined to Siberia, not only by ice, but by a land bridge also permitting the passage of wild life and of man (Eskimos and Red Indians) between the two continents. More than this, the weight of ice on some land surfaces, as in the Baltic, was so great as to tilt it downwards. Consequently, when the ice receded, though many land surfaces became flooded,

Table 2 *Chronology of the Glaciations During the Pleistocene Epoch*

Stages of glaciation		*(Years ago)*
	III	25,000
Last glaciation	II	72,000
	I	115,000
Last interglacial		
Penultimate glaciation	II	187,000
	I	230,000
Penultimate interglacial		
Ante-penultimate glaciation	II	435,000
	I	476,000
Ante-penultimate interglacial		
Early glaciation	II	550,000
	I	590,000
Villafranchian, when earliest worked stone tools are found		1,000,000

This table shows glacial chronology in west and central Europe and covers the million years of man. During the period of the Pleistocene, increasingly sophisticated stone tools were made.

others relieved of the weight of ice actually rose and ceased to be submerged.

The earliest traceable glacial episode in earth history[3] occurred in late Pre-Cambrian times, 600–700 million years ago. In the succeeding Palaeozoic era, there was a temperate climate everywhere, succeeded in later Palaeozoic times some 250 million years ago by glacial conditions particularly in the southern hemisphere. The last ice age, which has affected us in modern times so much, occurred during the Pleistocene period, during which modern man evolved. The Pleistocene glaciation consisted of four glacial epochs separated by three interstadials. The ice first encroached during the early glaciation some 590,000 years ago, but had receded 40,000 years later. The second glaciation occurred some 500,000 years ago, and lasted 25,000 years. The third glaciation started some 230,000 years ago and may have lasted 50,000 years. The last glaciation started some 115,000 years ago and the ice was retreating between 25,000 and 10,000 years ago. Typical arctic tundra still persists in Greenland and other northern areas. The changes that have occurred in the vegetation during the retreat of the ice have been determined with great exactitude by studies, particularly in Scandinavian countries. Not only has it been possible by geological studies to determine the limits of the ice and the glaciers and to assign dates to them, but by pollen studies it has been possible to ascertain the dates at which new vegetational successions followed the tundra. Much of the Sahara was still fertile and supporting nomadic herdsmen in early Egyptian times, and in parts plainly man's own efforts and those of his goats hastened the process of desiccation.

The end of the last glacial epoch is thus recent history. The floods that accompanied it are recorded in man's written history, and the myths and legends and taboos of an earlier period survive in cave paintings and in recollected superstitions and customs. Although the recession of the ice to its present limits seems to have occupied some 10,000 years, yet the climatic change which initiated it seems to have been quite sudden, and quite unconnected with any man-made activities, pollution of the atmosphere, and so on. The onset and retreat of glacial conditions are plainly due to natural phenomena, and one must suppose that the forces involved are far greater than any that man can command. A return of the ice would inevitably and inexorably drive man from his northern habitats, which he would be powerless to save. The point is stressed, because it is so seldom realised how altogether puny are the forces at man's command, even the most sophisticated, in relation to natural phenomena. No doubt man can destroy himself and temporarily pollute sections of the earth's environment; he can make little lasting impact.

Continental Drift

A true understanding of earth's past climates is, in fact, difficult to achieve. We see in tables compiled in accordance with the geological evidence that at times the whole earth has been warm and temperate even at the poles. Oil is found in Alaska and coal in Spitzbergen. How can this be, when the sun never shines during the Arctic and Antarctic winters? Surely these areas must always have been bitterly cold? Part of the explanation, though not the whole of it, is continental drift. That is to say that land masses were not always situated where they are today.

In brief, the land masses of the earth are not solidly anchored, but float on a layer of the earth's core which is known as magma and is in fact molten rock. We have seen that under great weights of ice, the land surfaces can be tilted and recover when the ice melts. It follows that continents can drift and in the course of time can occupy different positions in relation to each other and in relation to the poles and Equator. Originally, eastern America and western Europe and Africa were linked by their continental shelves, and Antarctica, Australia and India adjoined South Africa. There then existed one large land mass, Pangaea, with shallow seas scattered over it. This land mass was split by rifting, separating the Americas from Africa and Europe, but leaving the great land mass of Australia, Antarctica, South Africa and India, which is called Gondwanaland. This drifted towards the Equator, but then also split, India joining the Asian land mass and Antarctica moving to the South Pole.

The evidence for such drifting is too strong to be controverted, and the previous fusion of the continents is further attested by the distribution of animal life, for instance the presence of marsupials, such as opossums, in South America and Australasia but nowhere else. Thus the Antarctic land mass must at some time have been much nearer the Equator and, if we could move the 2 – 3 mile depth of ice which now covers it, we should find fossil evidence of primitive mammal life, particularly of marsupials. Land masses, therefore, have not always been where they are now, and coal remains at the North Pole would not necessarily indicate a tropical climate in polar regions. Evidence of the great tectonic movements which accompanied the separation of the continents is still to be seen in our great rift valleys and mountain ranges, such as the Great Rift Valley in East Africa, the Himalayas, the Alps and the Andes. Much of this rifting still continued into the times when primitive man was making stone tools.

I may have seemed to have strayed from my initial subject, the ocean/atmosphere system. In fact, this is not so. The great ice ages, the drifting of land masses, and the position of the continents must inevitably affect earth's climates and the exchange of water vapour and other gases between the oceans and the land surfaces. Today, these upheavals are largely over and the earth's surface seems stable and placid. Nevertheless, earth's crust contains well-known areas of instability, such as the well-known San Andreas fault through California. Who knows when some other great upheaval might not occur along lines of weakness, with consequent climatic changes from alterations of ocean/land relationships? Catastrophic rains or a new ice age could well upset the nice balances which decree our pleasant climate.

Ecological Cycles

The flow of nutrients through living systems is also usually studied by means of 'cycles', and such cycles can be compiled for virtually all the elements that are of importance in the construction of living matter and for its metabolic processes. The most important of these are nitrogen, carbon, sulphur and phosphorus. However, before passing to the cycles,

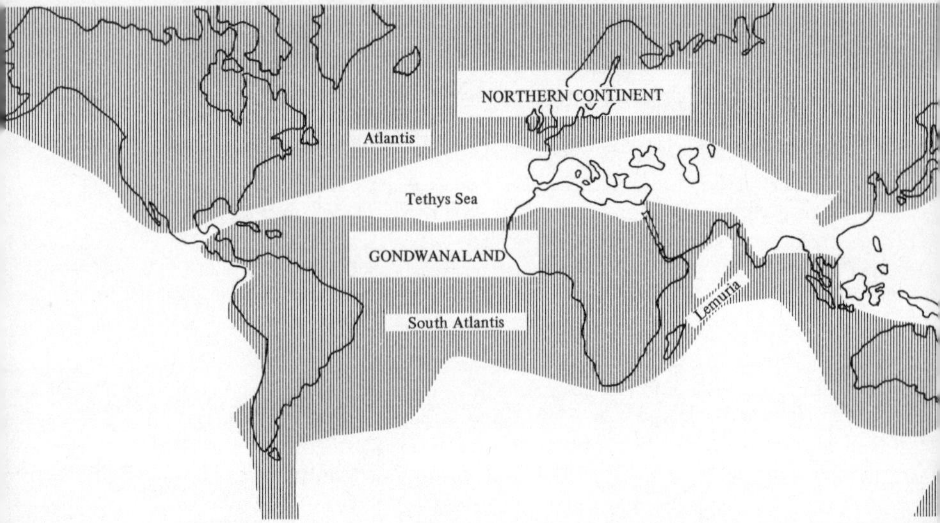

from Cloudsley Thompson *Terrestrial Environments*, p.4: Croom Helm, London, 1975

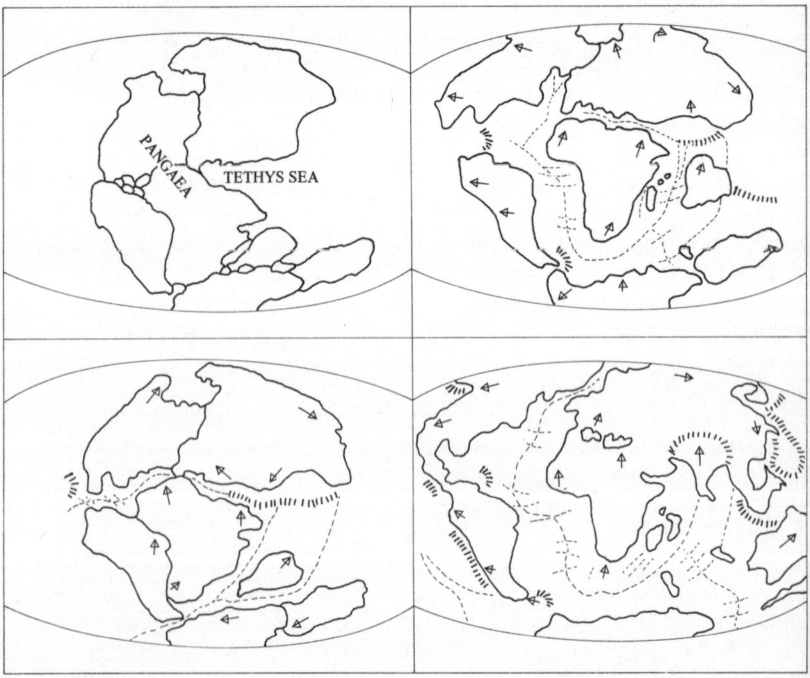

let us examine how these elements fit into living systems.

As a result of photosynthesis, carbon dioxide (CO_2) and water (H_2O) are combined to form carbohydrates. One such is the six carbon sugar glucose, of which the empirical formula is $C_6\ H_{12}\ O_6$. The structure is simple:

$$\begin{array}{cccccc} \mathrm{OH} & \mathrm{OH} & \mathrm{OH} & \mathrm{OH} & \mathrm{OH} & \mathrm{OH} \\ | & | & | & | & | & | \\ \mathrm{C}- & \mathrm{C}- & \mathrm{C}- & \mathrm{C}- & \mathrm{C}- & \mathrm{C} \\ | & | & | & | & | & | \\ \mathrm{H} & \mathrm{H} & \mathrm{H} & \mathrm{H} & \mathrm{H} & \mathrm{H} \end{array}$$

There could plainly be variations on this theme; one such is galactose, the sugar of milk; another is fructose or fruit sugar. These are known as monosaccharides. When two monosaccharides are linked, a disaccharide is formed; thus glucose linked to fructose makes sucrose or cane sugar, the sugar we use in our tea or coffee. When further combined, sugars form polysaccharides; such are starch and the woody tissues of plants.

The second important range of biochemicals is that of the fats. These are built from fatty acids, of which the empirical formula is $CH_3\ CH_2\ CH_2 - COOH$. You can insert as many CH_2s as you like making a whole range of fatty acids with different names; amongst the more simple are acetic (the acid of vinegar), and butyric (the acid of butter). Again the arrangement is simple:

$$\begin{array}{ccccccccc} & & \mathrm{H} & & \mathrm{H} & & \mathrm{H} & & \mathrm{H} & \\ & & | & & | & & | & & | & \\ \mathrm{H} & - & \mathrm{C} & - & \mathrm{C} & - & \mathrm{C} & - & \mathrm{C} & - \mathrm{COOH} \\ & & | & & | & & | & & | & \\ & & \mathrm{H} & & \mathrm{H} & & \mathrm{H} & & \mathrm{H} & \end{array}$$

When esterified (linked to alcohols), they form fats. In biological systems, they are linked to the alcohol glycerol to form vegetable or animal fats. In this form, fats are high sources of energy, and are the depot fats of animals. However, fats also occur in the unsaturated forms, in which they are of the greatest importance in body structure being known as essential fatty acids. In this form, some of the hydrogen atoms are missing, so that the compound is chemically unsatisfied and, therefore, capable of more complex linkages. The carbons in this case are linked by double bands where the hydrogens are missing; there may be only one double band or a number thus:

$$CH_3 - CH_2 - CH = CH - CH_2 - CH_2 - CH_2 - COOH$$

These unsaturated fats make the liquid vegetable oils such as olive oil or corn oil. They enter into the composition of all cells and in complex chemical form linked with phosphorus and other compounds form the white sheaths of nerve tissue. If not present in the diet in adequate amounts and the correct formulation, all sorts of serious results follow.

An amino acid is formed by inclusion of nitrogen in a fatty acid formula thus: $CH_3 . CH_2NH_2 . COOH$, to take the simplest. These acids are amphoteric, that is they can act as acids or alkalis, the carboxy group (COOH) being acid and the amino group (NH) being alkaline; they thus have a wide range of possible groupings. When combined, they form polypeptides; polypeptides combine to form proteoses; and proteoses combine to form proteins. There are twenty amino acids which occur in living systems, of which ten are dietary essentials. The proteins are the essential building components of the body – muscle, connective tissue, skin, horn and nails.

Apart from their importance as structural elements, many of the proteins play a part in metabolic activities as enzymes, without which the biological machine could not function. Amongst the most important are those proteins constructed of amino acids which contain sulphur in the form of a sulphydryl (SH) group. The sulphydryl group acts as a ready electron acceptor or donor, and is therefore necessary for energy pathways. Sulphur is, therefore, essential to living systems, and, being a somewhat rare element, its return to the pool by the sulphur cycle is of great importance.

Phosphorus also is an essential component of the energy system, having a number of functions. It is of particular importance, as we have already seen, because it is a component of the great energy transducer adenosine triphosphate (ATP). It is also an essential component of bones, nerves and other tissues.

A great many other elements are, of course, required both as structural elements in a living body and in metabolic processes. All of them must be present in the diet or mineral additives, or in salt or salt licks. However, the importance of ecological cycles resides to the greatest extent in the four mentioned, so we shall study them in these, to show the principle of what happens.

Oxygen Cycle

Let us first look at oxygen, which we have already encountered as a pollutant of the atmosphere in early times, when life first appeared on earth. Even today, oxygen is not a *natural* constituent of the atmosphere. By this, I mean that, if earth should again be bereft of its vegetation, oxygen would again disappear from the atmosphere. Expressed in another way, almost all the atmospheric oxygen is deposited there by plant life. Some 70 per cent of the oxygen is derived from the marine plankton, that is by the one-celled plant life of the oceans. The remaining 30 per cent is derived from plant life on land, chiefly forest trees; grass and other soil covers are less efficient.

It is a tribute to the photosynthetic efficiency of these plants that the level of atmospheric oxygen remains so remarkably stable. Nevertheless, fears have been expressed by a number of competent scientists that this will not always be so, for three main reasons. First, man's requirements of oxygen have increased to such an extent that oxygen levels in the atmosphere over major cities already show a deficiency of some 6 per cent, compared with the surrounding countryside. This is partly due to the oxygen required for respiration and partly to oxygen used in burning fuel for heating, transport, factories and other purposes. Secondly, the actual removal of plant cover in cities, roads, airstrips and other areas, together with forest clearance, has limited the power of plants to replace oxygen from land surfaces. Thirdly, any serious reduction of marine phytoplankton could lead to a reduction of photosynthesis of such an extent as to threaten oxygen reserves. In limited areas, such reductions of phytoplankton have occurred as a result of water pollution by detergents and chemicals, such as DDT and dieldrin.[4] Furthermore, closed waters as in the Great Lakes have been rendered virtually sterile because of the introduction of nitrogen in the form of its oxides, nitrites and nitrates and of ammonia from sewage and agricultural fertilisers washed into rivers from the fields. We shall look at this further when we study the nitrogen cycles. There is no immediate threat to the phytoplankton on a worldwide scale, but evidently the indiscriminate discharge into the sea of substances that could harm it requires to be controlled.

The actual oxygen cycle is a simple one. Plants in photosynthesis build carbohydrates from atmospheric carbon dioxide and water; during the process, excess oxygen is released into the atmosphere. The oxygen in the atmosphere is used in the respiratory processes of living things, including plants, and in the burning of fuel. The end point of these processes is carbon dioxide or carbon monoxide, which is converted in the atmosphere to carbon dioxide. The atmospheric carbon dioxide is reused by plants.

Carbon Dioxide Cycle

The carbon dioxide cycle, also, is very simple. In spite of this, as we have seen, the CO_2 content of the atmosphere remains extremely constant over periods of millions of years. It has been claimed, as already stated, that there has been a small rise in recent years as a result of the burning of fossil fuels, and that from a greenhouse effect the average temperature of the northern hemisphere has risen by a few degrees. It is also argued that this association may be erroneous, or alternatively that it is offset because of an increase of dust and other solid particles, which have the effect of insulating the earth's surface and thus cooling it. Is is thus undecided whether the extinction of life will come because the earth will be an inferno, or whether it will become entirely glaciated!

At any rate, carbon as one of the basic raw materials of life is not in short supply. It is plentiful in the atmosphere as carbon dioxide, and in the seas as dissolved carbon dioxide, as sodium carbonate and bicarbonate, and both in sea or land combined with calcium in the form of chalk, limestone, marble, gypsum and so on. There is also a great deal of carbon still locked up in fossil fuels, such as coal, oil, peat and lignite. It is the relationship of the seas to carbon that maintains atmospheric carbon at constant levels. When these levels tend to rise, more carbon dioxide becomes dissolved in sea water, and when the level in sea water rises carbonic acid is formed and this is combined with sodium ions to form carbonate and bicarbonate of soda. The seas are so vast that the effect on the composition of sea water is negligible. However, there appears to be some time lag in the process, and excess of carbon dioxide excreted into the atmosphere is not immediately removed; hence the small present-day increase of atmospheric carbon dioxide. It is relevant to remark that fossil fuels are expected to last at present-day levels of consumption for no more than a hundred years or so, and there appears to be no reason why the system cannot absorb them without serious consequences.

The cycle, as said, is inherently simple. Carbon dioxide is added to the atmosphere by the exhalations of animals (and of plants during the hours of darkness), by the combustion of fuels, and by the decomposition of dead organic material by microbial action. It is abstracted from the atmosphere by plants in the course of photosynthesis. Any imbalance in the process is corrected by the effect of sea water.

Nitrogen Cycle

Unlike carbon, nitrogen is a commodity in short supply, though equally with carbon it is an essential building material of living things. It would be more accurate to say that it is in short supply in a form that is available to living things. Higher plants obtain their nitrogen through their roots, not directly from the air, in which of course it is abundant. Moreover, it is only available for use in the form of nitrates, the oxidised form containing three oxygen atoms. Thus a number of complicated reactions occur in the soil which prepare nitrogen for plant use, the sources being: (1) atmospheric; (2) from broken-down organic debris (humus); (3) from artificial fertilisers.

Farmers using modern methods find it easier and more profitable to supply their crops with nitrogen by way of artificial fertilisers. It is not yet clear whether such systems, though in the short run profitable, can be sustained indefinitely. They include a number of undesirable features which may in the long run be self-defeating. The systems include the intensive use of the so-called agricultural chemicals for control of plant parasites and pests. The result is a virtual sterilisation of the soil, which is used as an artificial culture medium for holding the plant nutrients found to be necessary. We have already seen that a natural fertile soil is a living entity, crawling with life of different kinds and an ecosystem of its own. The living matter in soils processes organic residues, making them available for reuse; it lightens the soil and aerates it, so that atmospheric nitrogen is made available, and it provides the crumb structure, which permits water percolation and prevents erosion by wind and water.

It is well known also that top dressings of nitrogen and other substances are carried away by surface run-off into streams, lakes and reservoirs. Thirty per cent or more of fertilisers are quite normally lost in this way. The excess nitrogen has a most unfortunate effect in enclosed or slow-moving waters, which is known as 'eutrophication'. The end result is to remove the oxygen from the water, making it incapable of supporting fish and other creatures requiring oxygen. It happens in this way. Water algae are avid for nitrogen, and when this is supplied in excess their numbers multiply enormously to give 'a bloom'; by photosynthesis, they increase the oxygen in the water, and this results in the production of very large numbers of small organisms, which feed on them. These organisms quickly use up all the oxygen with the result that they and all other oxygen users die and the waters become a polluted, stinking mass of decaying material. Recovery from this condition may take many years. As a result of agricultural run-off together with sewage deposition, Lake Michigan is now a dead lake and may have gone past the point of no return.

In nature and still, fortunately, in our gardens, nitrogen problems are handled according to ecological laws evolved over many millions of years. Organic material suffers decay as a result of microbial action. Nitrogenous material is converted by one set of organisms to nitrites (oxide of nitrogen with 2 0 atoms), thence by another set of organisms to nitrates (with 3 0 atoms) available for use by plants. In well-aerated soils, atmospheric nitrogen is 'fixed' by other soil-living organisms, including those that are found in root nodules of conifers and leguminous plants. At the same time, organic debris forms humus and a rich tilth suitable for plant growth is produced. Some of the decayed material is converted to ammonia, thence to the oxides of nitrogen, both of which find their way back to the atmosphere to form nitrogen gas.

Nitrogen, as nitrate, is rather a rare material and very expensive to produce by synthetic means from atmospheric nitrogen. The tragedy of the present position is the wasteful and harmful way in which this rare material is being used. Not only does this apply to its use as fertiliser. Sewage systems are equally to blame. It is madness, and should be on the social conscience of industrial man, that the nitrogenous residues of urban centres should be swept out to sea whence they are irrecoverable instead of being processed for recovery of vital materials.

Sulphur Cycle

Another rare material essential to life is sulphur, and this is being equally wastefully exploited. The cycle is basically rather simple. Sulphur, as we have seen, is important in metabolic processes and present in all organic debris, from which it is released during decomposition as H_2S (hydrogen sulphide). Provided oxygen is present, this gas becomes oxidised to form sulphate. In this form it is again assimilable by plants and no problems arise. In the absence of oxygen, as may occur in marshes, estuaries, or thick reeds, the H_2S accumulates, giving rise to foetid conditions with a strong smell of rotten eggs. In such situations, it is eventually used metabolically by anaerobic sulphur-loving bacteria and photosynthetic green and purple bacteria. Sulphur in the form of sulphur dioxide is an important element of the smoke arising from coal burning. In this form it is dangerous to health, particularly in fogs or smogs, and a serious atmospheric pollutant. The discharge of sulphur, into rivers or estuaries as well as into the atmosphere, cannot be afforded. The world's sulphur resources are limited, and in the form of sulphuric acid it is of commercial importance. Its recovery from sewage or smoke presents no great difficulty, and the continuing folly of polluting water and atmosphere with this scarce element is incomprehensible.

Like sulphur, phosphorus is of great importance in metabolic processes and left to itself recycles by way of absorption by plants from organic debris in the soil. However, there do exist in the world limited deposits of phosphate in rocks and so on, and farmers find it more economical to treat their land with phosphate fertilisers than to rely on natural cycles. Excessive run-off from farm lands, together with phosphate in sewage and derived from phosphate detergents, assists the eutrophication of waters, besides squandering an important resource. Phosphate has been an important element in the destruction of the ecosystems of Lakes Michigan and Superior.

Cycles can be constructed for other elements important in metabolism, such as iron, copper, iodine, manganese, and so on, but

Fig 22 CARBON CYCLE

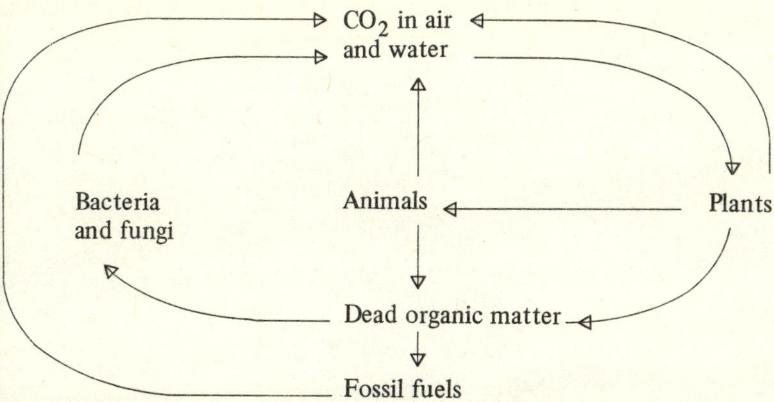

Fig 23 NITROGEN CYCLE

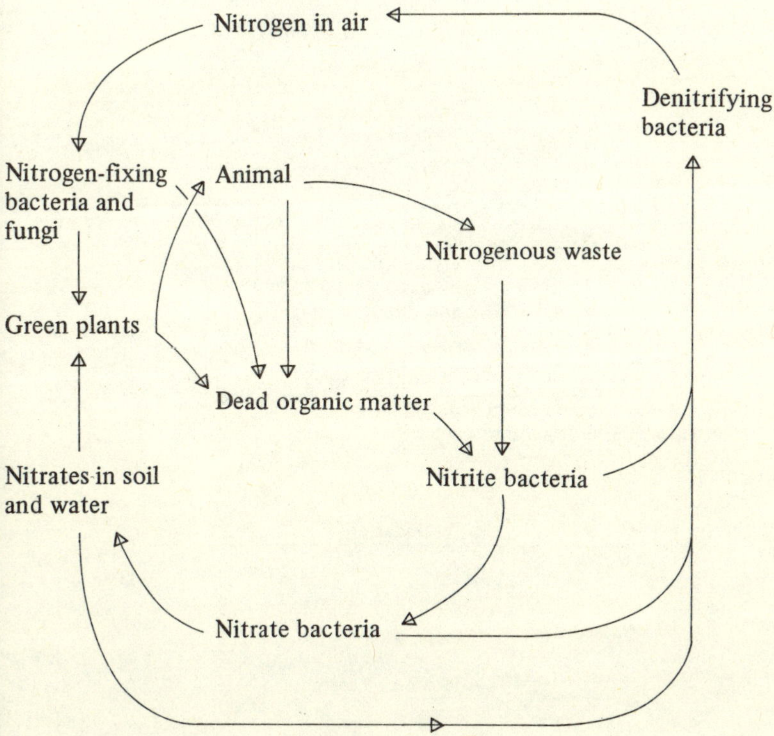

Fig 24 SULPHUR CYCLE

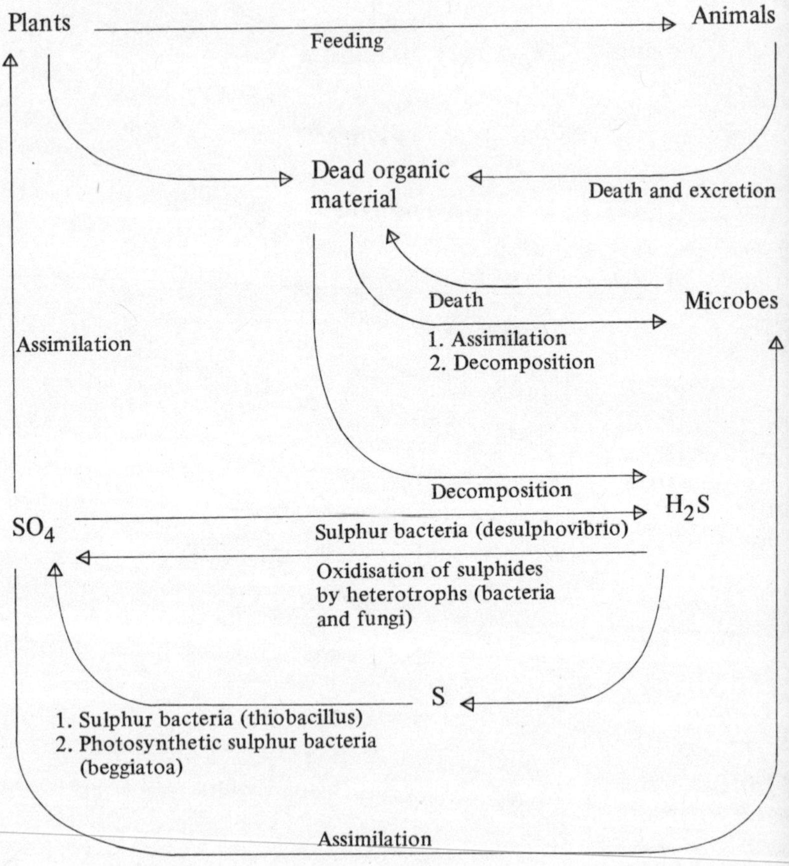

enough has been said to emphasise the importance of these cycles in nature's ecosystems and of the ways in which man is disrupting them. One of the most important problems of modern life is that of recovering some of these elements from waste materials and of returning them to the biological flow system. Mishandled, as now, these valuable materials become dangerous pollutants. Paradoxically, their recovery can be profitable.

We now pass in the next chapter to study the ways in which life forms interact with each other to form an ecological unit, the 'associes'.

NOTES

1 R.N.T -W- Fiennes, 'The Evolution and Colonisation of Habitats', in R.N.T -W- Fiennes, *Biology of Nutrition,* Pergamon Press, Oxford, 1972
2 G.H. Plass, 'Carbon Dioxide and Climate' in R. Ehrlich, J.P. Holden, and R.W. Holm, *Man and the Ecosphere,* W.H. Freeman & Co., San Francisco, 1971
3 E.C. Olsson, 'Climatic Change and its Influence on Life and Habitat', in Fiennes, op. cit.
4 G.M. Woodwell, 'Toxic Substances and Ecological Cycles', in Ehrlich *et al.,* op. cit.

6 THE INTERDEPENDENCE OF LIFE FORMS – THE 'ASSOCIES'

The ways of life and adaptation to terrestrial and aquatic environments of life forms are reviewed in subsequent volumes of this series.[1] Here it is necessary to survey the ways in which the members of a habitat interact with each other.

Modern works on ecology largely ignore the pioneer studies and theories put forward by the British zoologist Alfred Russell Wallace, who proposed the theory of natural selection at the same time as Charles Darwin gave his evidence of the mode of operation of evolution. Wallace's *Geographical Distribution of Animals,* published in 1876, was in fact the first major work on ecology, though that term had not then been invented. He pointed to the importance of natural and physico-geographical barriers in determining the distribution of plants and animals and to the results occurring when these barriers are removed. In particular, he drew attention to the effects of the long glacial epoch in dividing the world into faunal regions, each with a defined fauna and flora developed in isolation. However, the recession of the ice led to a redistribution of these regions and a tendency for a mixing of faunal types, with far-reaching ecological effects.

He puts so well some of the lessons this book attempts to expose, that it is worth our while to quote the eight principles he propounds:

> (1) If the dry land of the globe had been from the first continuous, and nowhere divided up by such boundaries as lofty mountain ranges, wide deserts, or arms of the sea, it seems probable that none of the larger groups (as *orders, tribes,* or families) would have a limited range; but, as is to some extent the case in tropical America east of the Andes, every such group would be represented over the whole area, by countless minute modifications of form adapted to local conditions.
>
> (2) One physical barrier would, however, even then exist; the hot equatorial zone would divide the faunas and floras of the colder regions of the northern and southern hemispheres from any chance of intermixture. This one barrier would be more effectual than it is now, since there would be no lofty mountain ranges to serve as a bridge for the partial interchange of northern and southern faunas.
>
> (3) If such a condition of the earth as here supposed continued for very long periods, we may conceive that the action and reaction of the various organisms on each other, combined with the influence of very slowly changing physical conditions, would result in an

almost perfect organic balance, which would be manifested by a great stability in the average numbers, the local range, and the peculiar characteristics of every species.

(4) Under such a condition of things it is not improbable that the total number of clearly differentiated specific forms might be much greater than it is now, though the number of generic and family types might perhaps be less; for dominant species would have had ample time to spread into every locality where they could exist, and would then become everywhere modified into forms best suited to the permanent local conditions.

(5) Now let us consider what would be the probable effect of the introduction of a barrier, cutting off a portion of this homogeneous and well-balanced world. Suppose, for instance, that a subsidence took place, cutting off by a wide arm of the sea a large and tolerably varied island. The first and most obvious result would be that the individuals of a number of species would be divided into two portions, while others, the limits of whose range agreed approximately with the line of subsidence, would exist in unimpaired numbers on the new island or on the mainland. But the species whose numbers were diminished and whose original area was also absolutely diminished by the portion now under the sea, would not be able to hold their ground against rival forms whose numbers were intact. Some would probably diminish and rapidly die out; others which produced favourable varieties, might be so modified by natural selection as to maintain their existence under a different form; and such changes would take place in varying modes on the two sides of the new strait.

(6) But the progress of these changes would necessarily affect the other species in contact with them. New phases would be opened in the economy of nature which many would struggle to obtain; and modification would go on in ever-widening circles and very long periods of time might be required to bring the whole again into a state of equilibrium.

(7) A new set of factors would in the meantime have come into play. The sinking of land and the influx of a large body of water could hardly take place without producing important climatic changes. The temperature, the winds, the rains, might all be affected, and more or less changed in duration and amount. This would lead to a quite distinct movement in the organic world. Vegetation would certainly be considerably affected, and through this the insect tribes. We have seen how closely the life of the higher animals is often bound up with that of insects; and thus a set of changes might arise that would modify the numerical proportions, and even the forms and habits of a great number of

species, would completely exterminate some, and raise others from a subordinate to a dominant position. And all these changes would occur differently on opposite sides of the strait, since the insular climate could not fail to differ considerably from that of the continent.

(8) But the two sets of changes, as above indicated, produced by different modes of action of the same primary cause, would act and react on each other; and thus lead to such a far spreading disturbance of the organic equilibrium as ultimately perhaps to affect in one way or another, every form of life upon the earth.

He subsequently classified the faunal regions of the earth, as determined in the glacial epoch, as under:

Regions

Neogaea	(Neotropical	Austral zone	Notogaea
	(Nearctic	Boreal zone)	
	Palaearctic)	)	Arctogaea
	Ethiopian)	Palaeotropical)	
Palaeogaea	Oriental)	zone)	
	Australian	Austral zone	Notogaea

The disruption of these zones since the end of the glacial epoch had depended first on climatic change; secondly, on the changed ecology of man, itself a result of climatic change. It is this that must now be studied, but first we must look at some facts of ecological stability, interference with which has led to the kind of instability that Wallace forecast. The main changes to be expected with climatic shift will be these:

1. Flora and fauna species may be no longer adapted to the new environment and new unoccupied ecological 'niches' may become available for colonisation. These vacancies will be filled either by adaptation – by natural selection – of the pre-existing life forms, or by invasion of the habitat by life forms from elsewhere. In practice, the latter has been predominant during the human era, and favoured by human activities. The results of invasion may be far-reaching and of long duration, as Wallace predicted.

2. New species invading the habitat may thrive to such an extent as to cause its deterioration. Instances are the introduction of coypu and mink into Britain, the grey squirrel into Britain, the rabbit into Australia, and goats into areas such as the Sahara fringes. The depredations of such animals may have very serious consequences leading to a complete change of the character of the habitat. For instance, goats – and cattle for that matter – will devour tree seedlings,

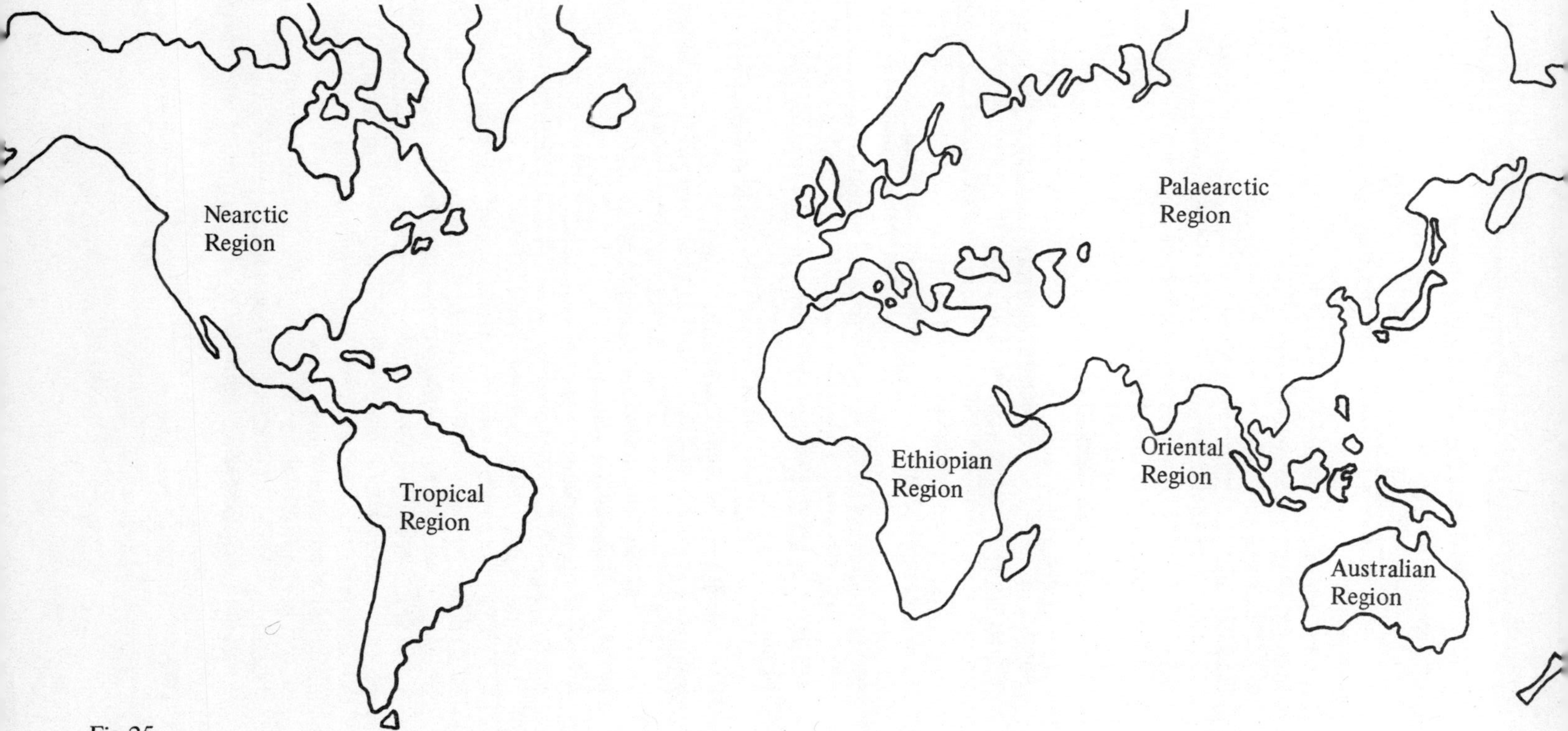
Nearctic Region
Tropical Region
Palaearctic Region
Ethiopian Region
Oriental Region
Australian Region

Fig 25

so that a forested habitat will become progressively grassland and – as has happened in areas of the Sahara – desert. Climatic changes can follow with desiccation, erosion, a fall in the water table and so on. The destruction of water-shed forests by man himself constitutes just such a challenge to the environment. The replacement of pre-existing species, of course, may lead to their extinction, and a great many animal species have indeed become extinct since the last glacial epoch.

3. The situation has been made worse by man's activities. This is especially the case where major predators such as wolves are exterminated. Their removal permits the multiplication of prey species to such an extent that natural forage becomes scarce or crops are destroyed. Man will then exterminate the culprits and the fauna of the land becomes the poorer. An example of this is seen in Britain, where a number of species of deer have been reintroduced. They have multiplied so fast as to be a nuisance and their numbers are controlled by periodical shooting of the excess population. A further instance of the results of man's interference is seen where, in order to preserve species threatened by the spread of human habitation, game parks are created. These parks favour the increase of static species but are inimical to those, such as elephants, which normally are seasonally migratory. So either the migratory species perform poorly in competition with the static, or alternatively, in order to sustain themselves on the restricted habitat, they virtually eat everything in sight and the habitat is destroyed. In this case again, man must intervene to control their numbers.

4. The invasion of a habitat, altered as a result of climatic change, is not made only by animals and birds. Insects, which carry disease, infiltrate also, and new parasites and disease germs of the invading species are also introduced. Even where an existing species is able to compete successfully, it may be weakened by the natural parasites of the invading species and succumb to the challenge. The disastrous decline of the red squirrel population in Britain in face of the grey squirrel is believed to have been as much due to its susceptibility to grey squirrel diseases as to direct competition.

5. Once the numbers of a threatened species fall below a certain limit, there occurs a point of no return from which revival is almost impossible. Above a certain lower limit of numbers, the population will thrive and in favourable circumstances numbers will recover. Below this point, the stock deteriorates, numbers diminish still more and the stock is likely to be extinguished. The reasons for this are not fully understood, but genetic defects incline to appear and to be perpetuated so that the constitution of the population is undermined. This occurs with human populations also, as is seen amongst such isolated communities as those of Tristan da Cunha. Even amongst such relatively

large populations as the whites of South Africa, unfavourable results from the perpetuation of genetic defects can be traced and they have been important in recent years in rural communities in European countries such as Britain, where the people were much inbred.

6. Ecological problems have also been intensified both in the human race itself and in the farm animals and crops raised, because of monoculture. Man's agriculture demands that crops be intensively cultivated, whether cotton, wheat, fruit, maize or anything else. This is directly opposed to the dictates of ecology, which demand diversity. In this way, while the desired crop is raised, the total potential biomass of a region is never realised. Moreover, parasites, pests and diseases are enabled to multiply in a geometrical progression and must be controlled if the crop is to be raised successfully. Hence the use of chemicals, bird scarers, and other devices to control these pests. It is not only with crops that these problems arise, because farm animals, particularly poultry (chickens and turkeys) intensively reared, are at especial risk to diseases and parasites, and these must be controlled by drugs and vaccines. Urbanised man himself is in the same situation, and epidemic diseases run quickly through crowded populations.

Once ecosystems are disturbed, whether by natural causes such as climatic change, or by invasion of life forms from outside, or by interference by man, far-reaching effects are the result, and these may be long lasting and deleterious. So long as an ecosystem, developed over thousands of years, is undisturbed, its numbers are self-adjusting. The herbivores are kept in check by the predators; the predators only kill for food not for the lust of killing; and the parasites and disease-producing organisms have become mutually adjusted with their hosts, so that serious epidemic diseases do not make inroads on the denizens of the habitat. The members of the ecosystem adjust their numbers by various devices, such as the well-known territorial system. In dry and cold seasons, those creatures which cannot find sustenance survive by a number of devices such as aestivation or hibernation, migration as with many bird species, or in the case of insects and annual plants in resting stages represented by fertile eggs or seeds.

Interdependence of Life Forms

Here we are concerned mainly with those rules of ecology which may be overthrown with undesirable results, and we need to study the mutual dependence of life forms on each other, that is commensalism, symbiosis, and their extension, namely disease.[2]

No forms of life can exist in isolation. This must be sufficiently clear from what has already been written. The stability of the ecosystems depends entirely on that of its members and their relationships with each other; on the organisms which promote decay, and thus ensure recycling of the elements, as much as on the primary and secondary producers. Throughout nature, there occur also innumerable instances of living things which combine together to make their living, and which could not flourish so successfully in any other way. This is known as 'symbiosis'. The most frequently quoted example is that of the lichens, seen everywhere on stone walls, etc., as pioneer life forms on surfaces which other plants are unable to colonise. By their activities, an arid and uninviting surface can be colonised and gradually eroded by the action of carbonic acid, so that crevices are formed to hold debris to become a sort of soil which secondary pioneers can invade.

Symbiosis is particularly prevalent in life forms on coral reefs. In such situations, corals, anemones, sea worms and even some protozoa contain chlorophyllous phytoplankton, known as zooxanthellae, in their actual bodies, and it has been suggested that in this situation they are responsible for much of the primary production on coral reefs.[3]

Now, mammalian herbivores in terrestrial situations are descended from insectivorous or carnivorous ancestors. Although they live on plant material, their digestive systems are not able to split cellulose and other higher polysaccharides, such as woody materials, lignin, and so on, which they consume. These functions are performed for them by bacteria and other micro-organisms in capacious fermentation compartments developed in their digestive systems. The most complex of these are the multiple stomachs of the ruminants, known as the rumen, the reticulum, the omasum and the abomasum. A similar development has occurred in the hippopotamus and the leaf-eating colobine monkeys. In horses, and other herbivorous animals, the fermentation chamber has been developed in a capacious large bowel, in particular the caecum. If the micro-organisms responsible are prevented from developing in the rumen of a calf, or are suppressed by antibiotic treatment, the animal cannot digest its food and will die. Some of these micro-organisms, furthermore, synthesise vitamins, particularly of the B group, and if eliminated by antibiotics vitamin deficiencies occur.

Apart from these beneficial organisms, all wild-living creatures harbour a variety of lesser creatures within their bodies, worms, protozoa, bacteria and possibly even viruses, which while doing them no harm do not positively contribute to their well-being. Such are known as 'commensals'. The commensals have a dual importance. First, in times of stress, illness, or malnutrition, they are inclined to become out of control and to cause symptoms of disease. Secondly, while harmless as a rule to the primary host, they may be dangerous disease-causing organisms to a secondary host, particularly one of a related but distinct species. In the first sense, their activities are probably beneficial to the tribe as a whole, because they tend to cause the deaths of weaker members not well attuned to them or to cause epidemic deaths when the group is over-populated, malnourished, or stressed, so helping to reduce numbers to a more favourable level. In the second sense, they only become important in circumstances of new introductions of animals to an ecosystem, or where an ecosystem is subject to change as discussed above.

A stable population, from tens of thousands of years of acclimatisation, is normally in balance with its parasites, and factors of disease are relatively unimportant. The heavy toll taken by disease both in human and animal populations in present days is the result of the disturbance of naturally developed ecosystems, and the introduction of strange parasites and diseases on the one hand, and of strange animals, including man, to new areas on the other. Monocultures, both human and other, favour the spread of these diseases. In addition, any activity which alters the environment may permit parasites, formerly under natural control, to gain ascendancy. The spread of Dutch elm disease is attributed to the destruction of hedgerows to permit combine harvesting. The natural control of the beetle which causes the disease is by woodpeckers, which delve into the bark for them and eat them. The woodpeckers are threatened by the removal of their hedgerow habitat.

The dangers of unaccustomed diseases in related species is shown by many instances which could be quoted, as for instance between man and other primates.[4] The best known of these is that of myxomatosis.[5] Myxomatosis is a naturally occurring disease in the American cotton-tail rabbits, in which it causes transitory skin tumours. When European rabbits become infected, it is virulent enough to cause the death of more than 98 per cent. When infected rabbits were introduced deliberately into wild rabbit populations in Australia and Britain, it almost led to their extermination, but not quite. As is well known, the rabbits are returning and the death rate from myxomatosis has diminished, although the disease still serves to limit their numbers. Many diseases, too, which are natural to monkeys are very dangerous indeed to human beings, and the greatest care has to be taken in handling them. The well-known disease caused by B. virus, one of the herpes group of viruses, is a case in point. While infection of humans is fortunately rather rare, only one patient so affected has survived and he only with his mental faculties permanently impaired. Some of the natural herpes viruses of monkeys are also very dangerous to other monkeys, and in some they cause a deadly form of cancer. This, too, is a point of great importance, because it is suspected that some human cancers are caused by viruses, including the leukaemias.

Table 3 *Summary of Geological Calendar with Approximate Ages in Millions of Years*

	ERA	PERIODS		AGE
			Epochs	
PHANEROZOIC	CENOZOIC	Quarternary	Recent	
			Pleistocene	
				2.5
		Tertiary	Pliocene	
			Miocene	
			Oligocene	
			Eocene	50
			Paleocene	
	MESO-ZOIC	Cretaceous		100
		Jurassic		150
		Triassic		200
	PALEOZOIC	Permian		250
		Pennsylvanian = L. Carboniferous		300
		Mississippian = E. Carboniferous		350
		Devonian		400
		Silurian		
		Ordovician		450
				500
		Cambrian		550
	PROT-ERO-ZOIC			600
		Metazoans		
CRYPTOZOIC	ARCHEOZOIC	Fungi		2300
		Blue-green Algae		2700
		First Traces of Life (Bacteria)		3300
		Origin of the Earth		

Impact of the Ice Age

The first major cause of ecological unsettlement, then, operative over the past 20,000 years, has been the climatic change associated with the recession of the last ice age (or glacial epoch as it is more correctly called). This led to fundamental changes of the environment and wide-scale invasion of habitats by strange life forms, and a mixing of life forms unaccustomed to each other. The second and more important factor was the change in the behaviour patterns of man, which were also promoted by this same climatic change. Let us see first how the recession of the ice has affected world ecology.

In Table 3 we can see a geological calendar of earth history. The period with which we are concerned is that known as the Cenozoic era, during which the main groups of mammals were evolved and became dominant over some 50 million years. During this time, the climate was warm and temperate; some 3 million years ago at the end of the Pliocene epoch, the last great ice age struck the earth. The history of the Pleistocene glaciations is shown in Table 4, related to the Palaeolithic stone cultures of man. The recession of the last glaciation has been plotted with a great deal of accuracy in Scandinavian countries, by recovering plant pollens from peat bogs and ascertaining the vegetational types present at different ages. This is shown in Table 4 which shows the world wide distribution of climates during the past 600 million years. The effect of the ice recession on the distribution of plants is sufficiently obvious from these tables. Nevertheless, these tables do not give a full picture of the extraordinary changes which occurred during the ice ages and their periodical recessions. The ice age deposits in Britain are revealed by the well-known boulder clays, which occur north of a line from the Thames to the Severn. In the lowest of these boulder clays are found the relics of typically ice-age animals, including extinct species of cold-loving elephants, and the so-called sabre-tooth tiger. These are overlain by riverine gravels indicating a climate probably warmer than today, and in these gravels are found the remains of animals such as the hippopotamus and straight-tusked elephants, which could only survive in warm climates. These gravels are again overlain by boulder clays containing the remains of cold climate animals, such as mammoth and woolly rhinoceros. These are again overlain by gravels containing the remains of hippopotamus, and above them more boulder clays, containing mammoth, reindeer, musk ox and arctic voles.

These changes have imposed severe strains on the life forms dependent on different types of climate. They have established themselves only to be again replaced by a different fauna; the new fauna in its turn has again been eliminated. It is hardly surprising that a goodly proportion of mammalian species have become extinct during these ages.

Table 4 *Climatic History of the Continents, Temperatures through the Ages (after Schwarzbach, 1963)*

	Arctic	*North America*	*Europe*	*Asia*	*Australia*	*Africa*	*South America*	*Antarctica*
Quaternary	Glaciations	Extensive glaciations	Extensive glaciations	Extensive glaciation in places	Glaciations in South	Pluvials	Glaciation especially in South	Glaciation
Tertiary and Mesozoic	Temperate to warm	Warm, arid in places	Warm, arid in places	Warm, arid in places	Warm	Warm, partly arid	Warm, partly arid	Warm at times
Younger Paleozoic	Warm,	Warm, partly wet, partly arid	Warm partly wet, partly arid	Partly wet, glaciations in India	Glaciations	Glaciations in South	Glaciations	?
Older Paleozoic	Warm	Warm	Warm	Partly warm	Partly warm	Cool at times at least in S.	Partly cool	?
Eo-Cambrian	Glaciations	Glaciations in East	Glaciations in North	Glaciations in East	Glaciations	Glaciations in South	Glaciations?	?

Of man alone, there is an unbroken record. He gradually improved his implements of stone and his hunting methods until the end of the last great glaciation, a matter of rather recent history.

What was happening to the fauna and to man in particular in more southern areas is less clear. However, it is known that during the northern ice ages, there was a great increase in rainfall in the tropics, known as Pluvials. These had two main effects. First, the great deserts, such as the Sahara and Kalahari, were well watered and fertile, supporting savannah-type vegetation and large populations of antelopes and gazelles. Within historical times the range of the African elephant, the hippopotamus and such animals as crocodiles extended to the southern shores of the Mediterranean. In the Sahara itself are numerous rock paintings of man hunting and herding where life today cannot be supported. Great tracts of the Sahara where no rain falls today are still overlain by a layer of fertile humus from plants that have long since disappeared. Secondly, the great equatorial forests extended far to the north of their present limits. The advance of the desert has undoubtedly been accelerated by man's pastoral activities and in particular the introduction of goats. Goats – and to a lesser extent cattle – browse on the vegetation and eat the tree seedlings, making it impossible for the vegetation to maintain a precarious foothold.

Man himself, of course, had trodden the hominoid or human path long before the ice descended, though the glacial climate was to influence his development in many and important ways. We shall in the next chapter trace the ways in which climatic changes have influenced the development of man and his way of life, and have led to environmental situations which affect the earth today. Meanwhile, an attempt has been made in this chapter to illustrate the dependence of life forms on each other and of all on the habitat. In spite of this, man's ancestors defied the ice ages when the other major components of the habitat were altered or eliminated, and even in those early days showed his capacity to defy the laws of nature, and survive, a capacity which he was to exploit in later ages with such profound effects.

NOTES

1 R.N.T -W - Fiennes, *Man, Nature and Disease,* Weidenfeld & Nicolson, London 1964
2 R.N.T -W - Fiennes, *Zoonoses of Primates,* Weidenfeld & Nicholson, London, 1967
3 F. Fenner and F.N. Ratcliffe, *Myxomatosis,* University Press, Cambridge, 1965

Further Reading

Fiennes, R.N.T.-W.-, 'The Evolution and Colonisation of Habitats', in R.N.T -W - Fiennes, *Biology of Nutrition,* Pergamon Press, Oxford, 1972

Healey, I.N., 'The Habitat, the Community and the Niche', in Fiennes, op.cit.

7 THE PROGRESSION OF LIFE AND THE EMERGENCE OF MAN

From Monkey to Man

The emergence of man as the dominant species of higher life on earth is a strange story depending on an ecological paradox. The environment made man, but subsequently man drastically altered the environment. The environment shaped man's ancestors as arboreal species, but it was environmental changes which deprived man's ancestors of their arboreal habitat, forcing them to a secondary adaptation to life on the ground.

During the great biological upheaval that occurred during the Eocene/Oligocene epochs, the main groups of mammals were differentiated much as they exist today. The carnivores were differentiated from the herbivores, though they both started from a common ancestor. At this time too, the ancestors of the primates took to the trees. Thick tropical forests encircled the earth on both sides of the Equator and a rich living (an unoccupied 'niche') was there for the taking by any animal group which became adapted to arboreal living. The primates lived – as many still do – on the abundant insect life, but also fed on fruits and nuts, lizards, small birds and eggs stolen from nests. Apart from a few snakes and eagles, there were no important predators, provided that the monkeys stayed in the trees. They became differentiated into genera and species, some living in the thicker central forests, some in the lower trees of the savannahs, some in the upper forest canopies, and some at lower levels.

Perhaps life in this idyllic environment was the origin of the Garden of Eden: certainly the serpent was the most important enemy. However, ecological laws as always dictated certain evolutionary and adaptive changes. Powerful teeth were not essential to defence, and the denizens of the jungle were left relatively defenceless. The sense of smell was less important than that of sight, so that monkeys developed keen sight at the expense of scent. The orbits were enlarged to accommodate larger eyes and came closer together for binocular vision. Manual dexterity was of importance for gripping the boughs and for picking and manipulating the food. Advantages lay in a troop moving in unison, so that social habits became developed and vocal communication was advanced, enabling the members of the troop to recognise each other, and even to pass messages when danger threatened or a good feeding ground was found. All these functions led to a greater development of that part of the brain, the cerebral cortex, which is associated with intelligence. The sense of sight in particular is intimately associated with this part of the brain, whereas the sense of smell is not.

Evolution occurred also in the ways in which monkeys moved through the trees. The most primitive forms, prosimians like the bushbabies, move by jumping from one vertical stem to another. The true monkeys, as opposed to the apes, jump on all fours from one horizontal branch to another. However, the true apes – the gibbons, orang utans, gorillas, and chimpanzees – developed a new mode of progression, known as 'brachiation'. They swing, Tarzan-like, at arm's length from one branch to another, using the resilience of the branches to assist the movement and quickly transferring their weight and balance. Where possible, the feet are used too, so that the weight is distributed amongst two or more branches instead of one. This mode of progression is far more efficient and lighter branches can support a greater weight of animal. Thus we find amongst the apes massive animals such as gorillas, chimpanzees and orang utans, whereas most of the true monkeys which still live in the trees are quite small. It is interesting that monkeys of the colobus group have also learned to brachiate; these animals live solely on vegetable fodder and carry round with them a heavy digestive apparatus, so that the better weight distribution when moving in the forest is advantageous to them.

In general, primates have adhered stubbornly to their ancestral habitats, where most of them are still to be found to this day. When the forest habitat has receded due to climatic change, the monkey populations have perished. Man's closest relatives, gorillas, chimpanzees, and orang utans, still stay in their forests, although their numbers are much reduced as forests are removed under human influence. Two groups of monkeys have to a large extent forsaken the forest habitat, the baboons and the patas monkeys. Both groups show modifications of structure and behaviour which have enabled them to survive away from the forest. The teeth are much enlarged for defence, especially in the males, and troops have improved social organisation adapted to the protection of the troop as a whole. The intelligence of baboons seems to be of a higher order than that of other monkeys, while the patas have developed long limbs and can move rapidly over ground surfaces. Their feeding habits have changed also, and they have learned to grub in the ground for roots, and the baboons sometimes hunt and kill animal food. Some Asian monkeys also spend a portion of their time on the ground and some have learned to find food in rivers and estuaries. So there are modern parallels for a phase of existence through which the human stock must have passed.

It has been proposed, particularly by the Dutch primatologist Dr A. Kortlandt[1] in his 'dehumanisation' theory, that surviving chimpanzees fall into two groups, those that are partially ground-living, and those that live solely in the trees. He believes that the ground-living groups may be related to proto-human stock, which began to be adapted to ground-living but reverted to the ancestral mode of life, whereas the proto-human stock stayed on the ground. He has shown by various experiments in Africa that the partially ground-living chimpanzees more readily use sticks as clubs and weapons and are more naturally aggressive to potential attackers.[2] Appearances apart, chimpanzees are so closely related to man judged by anatomical and many other criteria as to make this feasible. Moreover, as has been shown by Jane Goodall,[3] savannah chimpanzees hunt and devour antelopes, bushpigs, monkeys and domestic fowls, whereas forest chimpanzees are strictly vegetarian.

However that may be, it is certain that no group of apes or monkeys has deserted tree living for the ground unless forced to do so by ecological circumstances, that is because the climate changed and the forest habitat disappeared. So it must have been with man, and palaeontological evidence tells us that this was so. However, unlike the baboons and patas monkeys, man overcame the dangers and feeding difficulties of life on the ground in a different way. He did not develop the dangerous teeth of the baboons and patas. Instead, he used and subsequently made weapons of offence. He used sticks and stones and bones, and presumably anything else that came to his hand. He developed an improved social organisation, and perhaps most important of all he learned to communicate on a far higher plane, eventually developing language and speech. This demanded an increase in the size of the brain, improved powers of thought and eventually the ability to communicate thought and abstract ideas.

The present-day distribution of the great apes, with gorillas and chimpanzees in Africa and orang utans in Asia, shows that the great ape stock must have migrated between these far distant points at a time when the earth was humid and the great African forests were linked with the Asian. Man is supposed to have descended from a group of apes which lived in Miocene times; fossils of one type of these have been found in the Siwalik Hills in India, and it is known as *Ramapithecus;* the other was found in Kenya and given the name of *Kenyapithecus.* However, *Kenyapithecus* so closely resembles *Ramapithecus* that both are now included in the single genus of *Ramapithecus.* Man was presumably descended from the Kenya branch of his family and – as Kortlandt believes – from some form close to the chimpanzees. Claimants to human ancestry have been found also in lower Pliocene strata, but the fossil record from the upper Pliocene is deficient. However, in the lower (early) Pleistocene there emerges in south and east Africa a group of creatures known as *Australopithecus,* whose resemblance both to *Ramapithecus* and to man is very close. Some forms of *Australopithecus* are said by some to merge into the earliest fossils that can be definitely classified as *Homo, H. habilis. H. habilis* merges into a form of man, *H. erectus,* who was widespread from Africa to the Far East, and who is regarded as the direct ancestor of *Homo sapiens.* Both *Australopithecus* and *Homo habilis* are believed to have fashioned and used simple pebble tools. One group of *Australopithecus, A. robustus,* developed massive jaws and teeth, believed to have been associated with a hard vegetable diet. The less specialised lived by hunting small antelopes and other animals, whose bones they also used extensively as weapons. They also attacked and killed their own people and possibly indulged in cannibalism.

However, the position of *Australopithecus* in human ancestry has never been universally accepted since some eminent students of the subject maintain that anatomically he was too ape-like to be placed in the direct line of descent. Recently, discoveries by Richard Leakey in the Lake Rudolph area of Kenya lend support to those who hold this view. A fossil form of *Homo* with definite human characters and a larger brain case than *Australopithecus* has been found of a date – some three million years ago – which is older than that of most australopithecine remains. This discovery would suggest a common ancestry for *Homo* and *Australopithecus,* the latter being unsuccessful and becoming extinct.

The emergence of man from the apes is one of the most fascinating stories of ecology. In spite of close anatomical and phylogenetic resemblances, let us be clear about the great gulf which separates habitat-independent man from his cousins the apes, so habitat-dependent as to be in danger of extinction because of the decline of the forests. Anatomically man is distinguished by his upright gait, his manual adaptation and the size of his brain. In powers of thought and communication he is as far removed from the apes as the apes are from the dinosaurs. In spite of this, man's special powers must have been evolved over a very short period indeed, probably less than a million years. This development could not have occurred without a period of intensive selection, in which enormous numbers of the proto-human race failed to survive. In virtue of this, a race of man had emerged in northern areas during the Pleistocene, indistinguishable from ourselves and evidently with the capacity to invent and learn the higher technologies which are a feature of our culture. How did they get there and how did they live?

Early Human Society

As we have seen, the earliest apes which can claim human ancestry were present on earth in as widely separated habitats as Africa and India during the late Miocene and early Pliocene, when the climate was warm and temperate. Little is known of the upper (late) Pliocene. However, during the Pleistocene when most of Europe, Asia and North America were glaciated, human creatures appeared to have moved north into the glaciated areas and to have developed a highly satisfactory and profitable way of life. For a creature developed in the warm climate of Africa and having no natural defence against the cold, this was indeed a strange thing to do. It was plainly possible only because of high intelligence, and the ability to protect himself against the cold by making clothes and footwear and as a result of the discovery of fire. His hunting ventures too required skill and organisation, since he was hunting animals of considerable size such as mammoths and wild horses, and he needed to protect himself against large and dangerous predators, such as sabre-tooth tigers and the large arctic hyaenas.

In the earlier years of this era, man is best known from the stone implements he manufactured, of increasing sophistication as the years passed by. As the successions passed from the boulder clays of the glaciation to the gravels of the warmer interstadials, it becomes clear that man at this time was more numerous during the periods of cold. During this long period of 100,000 years, the tribes of the north evidently became accustomed to the arctic climate and this provided for them a habitat with which they became well associated.

The main type of vegetation in these areas was that called 'tundra'. This was an arctic or alpine form of vegetation which was low growing. It was very lush because of the numerous streams emanating from the ice cap which permitted rapid growth during the summer season. It supported enormous herds of grazing animals, which owing to the open nature of the country were easily hunted by an animal that had developed the skills of man. Indeed, Palaeolithic man constantly indulged in acts of slaughter, which to say the least of it were improvident, and he has been accused of being responsible for the extinction of the mammoth. Men appeared at that time to live in small groups. The main habitats were caves or rude huts built of mammoth bones and tusks, which were to the south of the ice fringes. The animals which were hunted were to a great extent waylaid on their migration routes as they moved north in summer and south in winter. Evidence suggests that parties of young men would travel north in the summer months and set up hunting camps by the sides of lakes. They would secure a bag of deer and others of the larger animals, and birds such as geese, and return to base with their provender when winter came. The meat could then be preserved beneath the snow or in whatever other type of deep freeze nature provided. They probably used sledges made of skins, and indeed skins and furs were much used for warm clothing.

By the end of this period, perhaps 20,000 years ago, men of modern type (the so-called Cro-Magnon race) had appeared in northern regions and had been living there in this kind of way for 100,000 years or so. No doubt we would still be living there in the same way, but for the climatic change which ended the Ice Age. Evidently during this time genetic forces had discriminated in favour of mental development. The more intelligent, at any rate, were equipped with the same mental powers as has enabled man to develop the technology of the space age. It would be inconceivable that these powers could have been developed in so short a time if it were not so. In art, too, the vigorous paintings of the well-known caves are in no way inferior to modern art, though less sophisticated. It is also evident that religion and magic played a great part in their lives. However, until the Ice Age ended, there was no incentive to create a change in a thoroughly satisfactory way of life. This way of life, however, was associated with the easy hunting provided by the open tundra. If the country became less open, life might become more difficult.

Just such a change occurred with the recession of the ice. As we have already seen, with this recession the country became forested. First pines invaded, then birch, and finally mixed forest. Man was not equipped to hunt in this kind of terrain, and had he been any other animal, he would have disappeared with the change of habitat. Some, of course, migrated further north, the ancestors of the Eskimos, and continued to live by hunting reindeer and seals. During the Mesolithic period, the Caucasian types who stayed behind had a very difficult time. In Scandinavia, their settlements – now permanent and not temporary – are found in the open land between forest and sea. Vast middens on the outskirts contain the relics of what they ate. There was much shellfish, and the bones of such forest animals as they could hunt or trap; they also gleaned nuts and other vegetable produce in the forest. The forests were held in awe and no doubt many legends and myths about the gods of the forests arose at this time. It is evident, however, that man was a shadow of his former self. His vigour had diminished and he was unable to recover in the face of this drastic climatic change. It was not to be expected that so resilient and intelligent a creature could be permanently subdued by these unfavourable circumstances, and the wind of change blew strongly over the land, a wind that was to bring profound ecological consequences to the whole face of the earth.

In the Near East, the Mesolithic period only lasted a few hundred years. In northern Europe it dragged on over 2,000 to 3,000 years. Its end was heralded by three revolutionary developments. The northern peoples domesticated the wolf, whose keen scent enabled it to hunt in the forests where man's keen eyesight was of little use. The fashion spread quickly to other parts of the world, and domestication of ox, pig, goat, sheep and horse followed. The second important development, which may have occurred in central Europe was the invention of the polished stone axe. These were long, narrow axes made, as the name implies, from hard polished stone, and well

Table 5 *The End of The Ice Age – climatic and vegetational fluctuations in Scandinavia since Palaeolithic times*

Glacial stage	*Date*	*Climate*	*Forest zones*	*Archaeology*
Post-glacial	400 B.C.	Sub-atlantic (cold, wet)	Beech	Early Iron Age
		Sub-boreal (warm, dry)	Mixed oak forest	Bronze Age Neolithic
	2,500 B.C.	Atlantic (warm, moist, oceanic)	Mixed oak forest	
	5,000 B.C.	Boreal (warm summers, cold winters; dry, continental)	Pine and hazel	Mesolithic
	6,800 B.C.	Pre-boreal (transitional)	Birch, pine, willow	
	7,900 B.C.			
Late-glacial	9,000 B.C.	Sub-arctic	Tundra	
		Warm oscillation	Birch and pine	
		Sub-arctic	Tundra	Upper Palaeolithic

hafted with wood. By means of these axes, the forest trees could be cut, and the forests no longer remained the masters. Not only this, but in the succeeding Neolithic era a whole range of carpenter's tools were developed from stone. There were saws made by fixing flint teeth in wooden frames, and adzes, even stone planes. For the first time wood could be worked. One of the most important developments from this was the building of river boats and ships that could go to sea, and the ground was prepared for the development of trade and commerce.

Meanwhile, the Near Eastern peoples had discovered that some

Fig 26 STONE AXE. Reconstructed by mounting the Neolithic flint head on a copy of a Neolithic haft preserved at the bottom of a bog. It was found that the full swing of the modern woodsman often chipped or broke the head. Using short, rapid strokes, the experimenters learned to fell trees more than a foot in diameter in thirty minutes. To fell small trees they chopped all the way around the trunk.

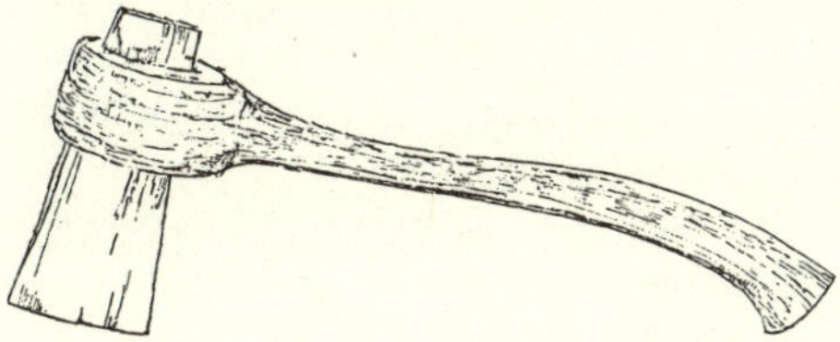

of the seeds they gleaned for food in these lean times could be made to grow if the ground was cleared and dug with a 'digging stick'. Particularly good results were obtained with the seeds planted in an area where the cut brushwood had been burnt. They took to clearing areas of woodland, burning the wood, digging and planting. Barley and wheat were the first crops to be planted. This primitive treatment of the soil necessitated frequent moves to new sites, and this form of shifting agriculture moved slowly across Europe until it eventually reached our northern friends with their domestic dogs. Meanwhile, peoples living along the great river deltas found that these seeds grew particularly well in the mud deposited during the annual floods. So developed the great river-based civilisations on the Nile, the Tigris and Euphrates, the Indus and the Yellow River in China. From these developments came the city states, commerce, banking, war, empires, and so to the recorded history of man.

While all this was happening, there was a group of peoples living in the vast steppe lands of eastern Europe and central Asia. These areas had not been invaded by forest; the chernozem soils supported a rich growth of valuable grazing grass, of which the dominant species is known as *Poa pratensis;* this grass was introduced to North America and is the famous Kentucky blue grass. This was all horse country and the wild tarpan horses were present in vast herds; in the colder north-eastern parts of this area in Siberia were the Mongolian wild horses (Przewalsky's Horses). The human tribes in this area were less affected by the change of climate and environment. They still lived in open terrain, and could pursue their traditional lives as nomads and hunters. They had been hunting wild horses from time immemorial for food. Although they were nomads, they were not to be left behind in the march of progress, and they were to develop in a way which has had and still has a profound influence on history. Their first contribution was to domesticate the horse and learn to ride it with the help of a crude bridle and bone bit, but no stirrups. They then turned to carpentry, invented the wheel, and made waggons, some of which were drawn by oxen and some by horses. Their crowning achievement was the construction of the two-wheeled, horse-drawn chariot, with light spoked wheels. They thus provided themselves with the two major military weapons which were to be supreme for centuries to come, cavalry and chariots. Nor did they hesitate to use these weapons; they continually invaded their richer neighbours to the south and east, and were a constant menace throughout the Seleucid and Roman Empires, up until the invasions from the east of Genghis Khan in the fourteenth century A.D.

It was in the late Neolithic era that commerce was really opened up through the then known world. The horse riders of the steppes took to the water as if born to it, and began to build excellent wooden ships, with which they and their emigré descendants traded throughout the Mediterranean, by way of the central European river systems (the 'Amber Route'), down the east coast of Europe from the Baltic into the Mediterranean, to the mouth of the Indus in India and down the east coast of Africa. There resulted a thousand years – 3,000 to 2,000 B.C. – of trade and prosperity in a climate that had so improved as to be sunnier and warmer than it is today. The evils of excess population growth, crowding, rivalry, greed, and continual warfare, were to follow in the succeeding Copper, Bronze and Iron Ages.[4]

The breathless bound of *Homo sapiens* from Ice Age nomads to agriculturists and traders and thence to the space age has all occurred within a few thousand years, perhaps six thousand. During the Ice Age, however, man no less than other groups of animals had been subject to the isolationist forces proposed by Alfred Russell Wallace. During the 100,000 years or so of this isolation, the races of man had diverged and showed clear racial characteristics developed in conformity with the habitats in which they lived. The races had diverged in physical characteristics such as colour, and in organising ability and intelligence. The peoples of the north had advanced furthest in technology; the darker skinned peoples of the south were left behind in this respect. To this day, peoples exist who are – or until very recently were – still living in a stone age culture, such as the Bushmen of South Africa, the Australian aborigines, peoples of New Guinea, and enclaves of peoples throughout Indonesia and the Indian subcontinent. To these we could add the Eskimos and the American Indians. The mixing of and intercourse between these races pose problems of the greatest difficulty in human ecology. As so often in ecology, it is the meek that inherit the earth. The apes that left the trees developed into man, and the human races forced into the bitter climate of the Ice Age became the dominant races of the earth. In the event of a total breakdown of western civilisation, who can say from what branch of the human race the next upsurge would come?

NOTES

1 A. Kortlandt, 'How do Chimpanzees use Weapons when Fighting Leopards?', *Yearbook* of the American Philosophical Society, 1965, pp. 327-32
2 A. Kortlandt and M. Kooij, 'Protohominid Behaviour in Primates (Preliminary Communication)', *Symp. Zool. Soc. London,* 10, pp.61-88 (1963)
3 J. Goodall, 'Feeding Behaviour of Wild Chimpanzees (A Preliminary Report)', *Symp. Zool. Soc. London,* 10, pp. 39-47 (1963)
4 G. Bibby, *Four Thousand Years Ago,* Collins, London, 1962

Further Reading

Bibby, G., *The Testimony of the Spade,* Fontana (Collins), London, 1962
Carrington, R., *A Million Years of Man,* Weidenfeld & Nicolson, London, 1963
Chiarelli, B., *Taxonomy and Phylogeny of Old World Primates, with References to the Origin of Man,* Rosenberg & Sellier, Torino, 1968
Cloudsley-Thompson, J., *The Zoology of Tropical Africa,* Weidenfeld & Nicolson, London, 1969
Fiennes, R.N.T -W - and Fiennes, A., *The Natural History of the Dog,* Weidenfeld & Nicolson, London 1968, and Natural History Press and Bonanza Books, N.Y., 1969
Goodall, Jane, *My Friends, the Wild Chimpanzees,* National Geographic Society, Washington, 1967
Goodall, Jane Lawick-, *In the Shadow of Man,* Collins, London, 1971
Napier, P., *Monkeys and Apes,* Hamlyn, London, 1970
Weiner, J., *A Natural History of Man,* Weidenfeld & Nicolson, London

8 MAN AND EARTH ECOLOGY

Man then is an arboreal animal, secondarily adapted to living on the ground. In this situation, he has also become adapted to be a highly organised social nomadic hunter in open situations such as tundra or prairies. By a third adaptation, he became a crops/pastoral agriculturist, by which many of the species still earn their living. By a fourth adaptation, he has become a highly urbanised animal, living in conditions of high population not matched elsewhere in the animal kingdom, except on a much smaller scale with ants, bees and termites.

As a predator during the Ice Age, he was a menance to his prey and other animals. If opportunity offered, he killed in excess of his needs – and still does. He selected for the kill the best animals in the herd, instead of the sick and the laggards usually taken by other predators such as wolves. Furthermore, he hunted not only for food, but also for skins and furs required to keep him warm, and in historical times at any rate he has hunted purely for sport and pleasure. He has been blamed for the extinction of the mammoth, of which he certainly killed enormous numbers. Even so, his responsibility for this has been questioned and many believe that the disappearance of the mammoth is to be attributed rather to sudden climatic change. While many other species became extinct at times during the Ice Age, this too is likely to be associated more with climatic change than with human depredation. Human numbers appear to have been quite small and as nomads the bands would be likely to shift their hunting grounds, and it is difficult to see how they could have exterminated whole species of animals.

At the end of the Ice Age, the position was radically changed. Animals that had previously been hunted were brought into captivity and reared as domestic stock. Conditions for the wild forebears of these animals deteriorated with the enclosure of land for farming or ranching, and because they were deliberately destroyed as being supposedly dangerous, or because they competed for pasture or stole the crops or killed the domestic animals. From Britain there have disappeared wolf, bear, beaver, polecat, aurochs, horse, elk, reindeer, and birds such as ospreys and golden eagles. Other species are endangered, and all over Europe, Asia, and North America the same story can be told. The rich fauna of Africa is threatened.

Vegetation-wise, the ability to fell the forests which started with the invention of the polished stone axe has had profound effects. Man's demand for timber, for building houses and ships, for making paper, and for burning, has been insatiable. Trees have been cut without adequate replanting programmes. They have been removed from vulnerable sites, such as watersheds, giving rise to diminution of rainfall, flooding, erosion and lowering the water table. Furthermore, animals inimical to the habitat have been introduced, such as rabbits and goats; tree seedlings have been cropped by them, so that natural regeneration has been prevented. All of the trouble cannot be laid at the door of modern man. Even in early dynastic times in Egypt the depredation of marginal but still useful lands had started. Plato lamented the disappearance of the trees from the hills of Attica, and the effect this had on water supplies. These hills are still denuded, and so have been the hills of Palestine since biblical times.

In the early days of shifting agriculture, probably little harm was done because of the necessity to leave the land to regenerate after a few seasons. In temperate regions too, well-planned agriculture, as practised in Celtic and Roman times and in countries such as Britain until recently, appears to have had no lasting ill-effects on the soil. Until recently too, the practice of shifting agriculture in tropical countries such as Africa has probably not been permanently deleterious. However, intensive agriculture in the tropics can easily lead to permanent loss of farm land unless very great care is taken. This occurs for two reasons. First, heavy tropical storms can easily remove thin topsoil which has taken centuries to be deposited (sheet erosion); they also cause great gullies which may be ten or twelve feet deep (gully erosion). Secondly, the effects of the hot tropical sun on the mineral-rich subsoil leads to 'laterisation', by which the soil is compounded into rock.

In so far as farming leads to monocultures, whether of crops or animals, it must also lead to the multiplication of pests, parasites and disease-causing organisms. The old-fashioned mixed farming with small fields and frequent hedges to harbour birds permitted these to be largely controlled by natural means. In intensive systems, these natural controls cannot operate, bird populations fall and pests – including bird pests such as pigeons – multiply. So the fields must be treated with chemicals and by burning the stubble. By this means, the soil-living ecosystem is also destroyed and such creatures as voles are deprived of their food and disappear. The ecological effects are profound. Already, wind erosion is threatening the soils of Norfolk in eastern England because of the removal of the hedges, and nobody knows what will be the long-term effects on soils deprived of their crumb structure and fed with massive top dressings of artificial fertiliser.

For North America, the Industrial Revolution gave rise to widespread erosion problems. The momentum of the revolution could only be sustained by large production of grain to feed the increasing factory population. Thus grain farming replaced cattle farming over huge tracts of the Middle West; the land was seriously mined and lost its fertility; dustbowl conditions were created. For England, too, this development was a tragedy, since the availability of imported grain permitted too rapid a growth of population and urbanisation at a time when little was known of hygiene and disease control, and large tracts of good countryside were destroyed. Aerial and water-borne pollution also started at this time. The effects are felt to this day.

Population Growth and Urbanisation

We can here take a quick look at the problems of population growth and urbanisation. We have already studied certain basic principles, such as:

1. That populations are maintained at optimum levels demanded by the ecosystem;

2. This maintenance is achieved by production of excess young and elimination of those least fitted to survive;

3. That mating and parenthood are denied to the young of the species until they have acquired a 'territory', that is a sufficient area in which to raise and feed young and provide protection;

4. That, if the population becomes excessive either because of a breeding rate that is too high or because conditions become unfavourable, certain glandular mechanisms operate to induce a state of 'stress', during which the breeding urge and fertility become diminished, aggressiveness towards other members of the community is enhanced, and susceptibility to disease is increased.

The most primitive human communities, such as the Kalahari bushmen and the Australian aborigines, were, and to a certain extent probably still are, subject to these natural laws. So, we must suppose, were all human tribes up to the time of the agricultural revolution. In subsequent human history, these natural laws have been progressively put aside, it would appear by deliberate intent, until in advanced human communities they are virtually inoperative, except in certain ways discussed below. However, let us examine these principles in turn in relation to human ecology.

The first principle, that populations are maintained at optimum levels for the ecosystem, has plainly been defied. Since the invention of agriculture, man has deliberately manipulated the environment and refused to conform to it. In many areas and at many times, he has exploited his habitat and caused permanent deterioration, to such an extent that it can no longer support his agriculture. When the damage is very extensive and deserts are formed, of course the human population disappears; where the damage falls short of this, the population tends to persist without great diminution of numbers at subsistence or below subsistence levels, malnourished and living in abject poverty. Disease is rife and frequent epidemics occur; famines and floods from destruction of ground cover recur at intervals. Still, the population maintains a high birth rate. However, expectation of life may be no more than thirty years, and infant mortality in particular is fantastically high. It is difficult to understand just why natural population-controlling mechanisms are not operative. Two possible causes may be suggested. First, such situations are not entirely unknown under natural conditions. Man could be regarded as an introduced species to a new environment. Under such circumstances, it often happens that the introduced species multiplies vigorously and destroys the environment, as has occurred with rabbits in Australia. The same has been happening with free-ranging animals, such as elephants, confined in game parks in Africa; their numbers increase and the confined habitat deteriorates. The problem is partly associated with the lack of predators in the new environment which will serve to limit the numbers of the introduced species. For whatever reason, man has developed to some extent into his own predator and limited numbers by wars of aggression, human sacrifice, cannibalism, and so on. The second cause may be associated with a deliberate reaction of an intelligent, reasoning creature in the face of a high mortality that threatens his race. Fertility cults are a feature of all religions, and exaggerated importance has been attached to an early and high rate of breeding. Fertility cults evidently first emerged in prehistoric, Palaeolithic times. No doubt in those times life was hazardous and high fertility was necessary to maintain the numbers of the community. Since man adopted a settled existence, the high fertility has been retained in spite of overpopulation and in defiance of nature's laws with resulting low standards of life and periodical mass exterminations from natural causes.

The second principle states overproduction of young and elimination of the least fit. Population depends on the balance between young produced and rate of survival. Animals that are especially vulnerable produce large numbers of young and survival is low. Small birds produce large clutches of eggs and less than half come to maturity or themselves breed. The extreme of this is seen with parasites, the larvae of which must find a new host, a somewhat fortuitous matter. A female tick produces thousands of eggs which develop into larvae, but the general tick population under normal circumstances does not increase. The same is true of intestinal worms and of protozoan parasites carried by ticks, and other vectors. The tendency of agricultural man also is to overproduce, but the balance is not redressed by comparable casualties. Epidemics and catastrophes apart, the main population-limiting factor in human history has been chronic ill health. This operates to the greatest extent on the very young and on the young and the mother at childbirth. In primitive, crowded communities neonatal deaths and deaths of mothers at childbirth are very high.

Until weaning, newborn children are protected to a great extent against diseases by immunity acquired from the mother in the milk. Thereafter, they are exposed to numerous diseases, of which malaria is probably the most important. A significant proportion of the juvenile population (up to 20 per cent) may succumb to this and other diseases, such as intestinal infections, pneumonia, and so on. If the population is malnourished, as is usually the case, the effect of disease is intensified and deaths may occur also from nutritional diseases, such as kwashiorkor due chiefly to lack of protein. Furthermore, intelligence and general performance during adult life will be impaired. Nevertheless, the birth rate remains at a level high enough to perpetuate these conditions, in spite of an early death rate in adult life, from diseases such as cholera and tuberculosis. The impact of more developed civilisation on peoples living in such conditions is disastrous. Attempts are made to improve food production and to reclaim exhausted land, but the impact of medical science is far more rapid in its effect. The killer diseases are brought under control and childbirth is made more safe. The result is population explosion on an unprecedented scale, with an exacerbation of pre-existing problems. Cataclysms and catastrophes affecting adult people become more severe. These are met by international relief programmes and shipments of surplus food from overseas.

The third principle that the young do not mate and breed until they have acquired a territory is no longer a rule of human ecology, though territorial instincts are strongly operative in us. Territory can belong either to a pair, as with a pair of birds which claim the right to the worm on the front lawn, or to a group of animals living in social conditions. Man appears to possess both instincts. A tribe will fiercely defend its own territory from invaders. Teenagers in cities, who form themselves into bands, seem to demarcate territories belonging to the different bands, and they usually respect each other's territories. On the other hand, it is the ambition of each young couple on marriage to own their own house, however humble, in which they can live together and raise their families. This instinct could be the saving grace of urbanised peoples, and many couples do obey the ancient instinct and defer marriage or babies until they have acquired their own home. This is made the more easy for them in Western societies, where birth control methods are readily available, and where the urgency of sexual instincts can be satisfied without the danger of unwanted children. However, this has not always been the case and is still not the case amongst peoples at a less advanced level, though evidence suggests that in many cases means of birth control would be welcomed. Amongst many peoples, the acquisition of a territory for young people is a virtual impossibility, but this does not act as a deterrent to breeding. Nevertheless, the instinct still exists and is a powerful force which may yet be important in human ecology in the future.

The fourth principle concerns the operation of 'stress' mechanisms in response to overcrowding. This is a somewhat complex problem, and will require rather more detailed study. Stress undoubtedly plays an important part in human ecology, although the human race as a whole resists stressing situations which would be intolerable for many wild animal species. Hans Selye, who has been a pioneer of the study and concepts of stress, regards stress as a general reaction to trauma in the same way as inflammation is a local reaction. The 'stressors' which cause the stress reaction can be either physical or psychological. In the present context, we are concerned with the psychological aspects, which have been studied extensively by animal ecologists in relation to population problems. The essential mechanism is fairly simple. The important pituitary gland acts as a coordinator of endocrine activity, by producing hormones that influence the other endocrine glands including the sex glands. It is situated at the base of the brain and its own activity is controlled by a nearby portion of the brain. When an animal is stressed, the pituitary sends a hormone to the adrenal gland, which stimulates its activity. In response, the adrenal gland produces more of its hormones which counter the stress situation, while the sex glands (and sex urges) are diminished. The results are a diminished urge to mate and reduced fertility – even abortion – while the animal enters a state of nervous tension, adopts unnatural habits and behaviour, and becomes more aggressive. If, as a result of continued or repeated stresses, the adrenal glands become exhausted, the animal will die since its secretions are essential to life.

The extent to which these stressing mechanisms arising from a crowded existence are operative in human ecology is a matter which must be considered. While man reacts atypically, the evidence is suggestive that urbanised human communities are indeed subject to the pressures of stress. At some time in their lives 10 to 20 per cent of men and women in urban areas receive treatment for neuroses of some kind, and an unreasonably high proportion of the population is permanently incapacitated or hospitalised for mental defects of one kind or another. It is unlikely in the extreme that any wild community, or primitive human community, could or would support so high a proportion of ill members. They would almost certainly fall prey to the natural predators of the species. However, man today has no natural predators, except his disease-causing parasites, and these, followed the many years of impotence against them, he more or less successfully controls. His principles are such that he preserves the weak and the unfit; and the drop-outs are, therefore, preserved as a burden on the community as a whole.

It might be thought that the long-term effect of this would be to put evolution and natural selection in reverse with a consequent genetic weakening of the community as a whole. However, Dr. Penrose, the British geneticist who has made a profound study of the subject, does not believe that this is necessarily the case.[1] He believes that, appearances notwithstanding, evolution is powerfully at work in the human species, but in a different way operating more on the unborn foetus than post-natally. He points out that more than 50 per cent of people in urban communities are 'genetically unfit'. This is to say that either they have no children themselves, or if they have children those children do not breed. Thus, in each generation there is a better than 50 per cent selection. The figures he gives are as under:

1. *Prenatal deaths*	15%
2. *Of the remainder there are*	
a. Stillborn	3%
b. Deaths of the newly born	2%
c. Deaths before maturity	3%
3. *Of the survivors*	
a. Those who do not marry	20%
b. These that remain childless	10%

The members of the population to breed most successfully are the middle-of-the-road ordinary people with a wide genetic constitution; those at the top or bottom of the brilliance league tend to be less productive. Therefore a wide genetic pool of capability is retained and those best adapted to their way of life are most likely to contribute the future generations. With so large a degree of infertility, there evidently does exist a widespread unwillingness or incapacity to breed. This, then, is just what we have seen to occur in wild animal populations that come into stress. The fact that, in spite of this, there may still be an upward movement of population numbers does not affect the argument, and it may well be that at some upper limit not yet reached a downward trend of population would result.

It can be argued then, without dogmatism, that urban crowding does indeed affect population trends amongst a section of the people. Are there other signs of stress effects to be seen in urban peoples? There are abundant signs, though it is impossible to prove that these signs are the effect of crowding or urbanisation. Let us consider them. The whole can perhaps be summed up in the term 'urban unrest', namely a feeling of insecurity and anxiety amidst the outward signs of affluence and prosperity. Here we may just list the types of unnatural behaviour that might be associated with stressing problems. First, there is the problem of young adolescents and their antisocial behaviour of whatever kind. Secondly, sexual freedoms and unnatural sexual habits. Thirdly, addiction to drugs and excess alcohol. Fourthly, the high level of lawlessness and crime.

Human history itself has been a story of alternating periods of vigour and decline. During the periods of vigour, wealth and population have increased. The periods of decline have been marked by signs of unrest leading to aggressiveness and wars, often accompanied by famine and pestilence; death and population decline follows. Here again, human history follows the ecological pattern of wildlife societies.

We can, then, discern in human society the operation of natural laws which govern pre-agricultural situations though they operate over much longer time intervals. Two new factors have been introduced into the ecological picture in recent times, namely the ability to control population increase by methods of family management, and the far-reaching advances that have been made in controlling disease. Amongst more advanced peoples, conditions of life have been so ameliorated that it is difficult to visualise what it was like to live even 100 years ago, and this amelioration has been achieved not only for the rich, but for the great mass of the people also. The expectation of life for the poor a century ago was under forty years, and even for the rich no more than fifty. Today, one can expect to live to seventy or more. Who would return to the good old days? Their dark, damp, and insanitary houses? Lack of water, drainage, sanitation? Long working hours without holidays? Deaths in infancy of up to half the children born? Twopenny gin and degradation?

Is the present the crest of the wave, before another period of decline? The general conditions of unrest would suggest it. An understanding of these problems may show the way to avoid it. Populations must be controlled at a point below that which leads to aggressiveness and instability, otherwise ecological laws will take over. Finally, the benefits of modern living must be extended to more backward peoples. This means, above all else, education, true education – not just literacy. It is useless to prolong by medical means the lives of people who are unable or unwilling to come into ecological balance with their environment.

NOTES

1 A.L. Penrose, *Biology of Mental Defect,* Sidgwick & Jackson, London, 3rd edn., 1963, and D.F. Roberts and G.A. Harrison (eds), *Natural Selection in Human Populations,* Pergamon Press, Oxford, 1969

Further Reading

Andrewartha, H.G. and Birch, E.C., *The Distribution and Abundance of Animals,* University of Chicago Press, Chicago, 1954
Bertram, Colin, *Adam's Brood,* Peter Davies, London, 1959
Harrison, G.A., Weiner, J.S., Tanner, J.M., and Bernicott, N.A., *Human Biology,* Clarendon Press, Oxford, 1964
Kormondy, E.J., *Concepts of Ecology,* Prentice-Hall, Englewood Cliffs, New Jersey, 1969
Lack, D., *The Natural Regulation of Animal Numbers,* Oxford University Press, Oxford, 1954
Selye, H., *The Story of the Adaptation Syndrome,* Aeta Inc., Montreal, 1954
Wynne-Edwards, U.C., *Animal Dispersion in Relation to Social Behaviour,* Oliver & Boyd, Edinburgh and London, 1962

9 THE GLOBAL ECOSYSTEM

Natural Laws of Ecology

It is a common trick in ecological studies to enclose a piece of land, protecting it from external influences and thus removing factors which may be deflecting the primary succession. In this way, one can study the regeneration processes and discover what vegetation would naturally develop. The influencing factors to be excluded may be various, such as rabbits, grazing by livestock, annual grass burning, and so on. The results of such experiments are often enlightening, and they may give useful information on the best ways to manage the land.

It is intriguing to consider what the results would be of a similar experiment, if man were totally removed from the earth. The main results can be deduced from simple observation. We all know how quickly weeds and undergrowth take over a neglected garden, until it becomes virtually impenetrable; how the ivy claws down walls and buildings, and kills introduced fruit and other trees. Even in derelict sites, such as bomb sites in a city like London, natural vegetation quickly returns. Around cities, a whole host of creatures, such as sparrows and pigeons, even foxes and hyaenas, which have learned to live by scavenging man's waste materials, would be forced to revert to more natural forms of life or disappear. Likewise, those that scavenge the grain fields would find life less easy. Within a few centuries most visible signs of human occupation are likely to have disappeared, and it would require an archaeologist of the future to rediscover and marvel at a lost civilisation.

In brief, the world's habitats would again be ruled by the natural laws of ecology. Where climate and soil favoured the development of forests, the land would revert to forest. In Britain, low-lying areas would again become covered by the thick tree growth which defeated our Palaeolithic ancestors. The moors and upland downs, where Neolithic man had his ridgeways and settlement areas, would revert to their natural vegetation. The great prairie lands of the world, such as those of the chernozem soils of North America and central Asia and Europe, would again become rich natural grazing lands. Man-made aerial pollution would disappear. Even the balance of carbon dioxide between atmosphere and ocean would be readjusted; any effects of releasing fossil carbon would be removed such as the supposed 'greenhouse' effect; and particulate matter released into the atmosphere by human activities would subside. If man's way of life has indeed affected the climate in these ways, then the climate would revert to that dictated by solar radiation, and whatever other influences determine the climate. Soil communities too would regenerate, and the removal of man-made pollution from inland waters, estuaries, and the oceans would enable the regeneration of aquatic creatures; fish would again teem in the seas, and in the inland waters.

On land surfaces, herbivorous animals would build up large populations within a short time. The bison would again cover the plains of North America. Horses would become feral, as did the mustangs in America during the last century, and take over the plains of Europe and Asia, possibly in America also. In forested areas, deer and elk would become numerous. In Africa, elephants and rhinoceroses would come into their own again, and vast herds of antelopes, gazelles and zebras would cover the plains. Predators might be slower to recover, especially in countries such as Britain where so many have been exterminated. However, the domestic dog has displayed in many instances its capacity to become feral and redevelop the properties of a social hunter. So we could expect the re-emergence of canine stocks resembling wolves, which would prey on the great herds of herbivores.

According to many of the more baneful prophets of doom, this is indeed what may happen. It is said that we are fouling our own nest to such an extent that recovery is already impossible. Some would go further and say that we will take all other living things on earth with us, destroyed by universally lethal radiation.

Some might think that man has made such a mess of the great opportunities offered him that from the earth's point of view his disappearance would be highly desirable, and that in course of time some other form of highly intelligent life would appear that would do better. However, would it? And if it did, would it not also pass through the same difficult stages that man is passing through at the present time? The point deserves some consideration both from the practical and the philosophical point of view.

Some of the more successful groups of animals are those which have become habitat-independent. Man is a case in point. Adapted to an arboreal habitat, he became secondarily adapted to life in open, terrestrial country and changed his diet from one that was mainly vegetarian to one that was partly carnivorous, and certainly omnivorous. The third refinement to appear on the ecological scene has been man's manipulation of the environment over the past 5,000 to 10,000 years, for the sake of his agriculture, and a further reversion to a primarily vegetarian diet.

These are surely inevitable ecological sequences, and should man disappear some other species will in the end appear and in its turn also attempt to manipulate the environment, going through similar stages of trial and error. Manipulation of the scene in the early stages was largely a matter of trial and error, but this has increasingly given way to more purposive procedures arising from observation of cause and effect and calculation of the results of actions taken. Hit and miss methods have given way to the scientific approach. It is not only man's early attempts to control the environment that can be described as hit and miss, the whole of unmanipulated ecology can be so described. Since the earth started, the development of its environments and of the life forms that go with them has been entirely fortuitous. The fact that a green and pleasant land came into being, and that the creatures living out their generations on it, without being aware of it probably derived enjoyment and pleasure from it, is beside the point. The point is that there was no creature on earth capable of mastering the laws which governed physical and biological processes, and in this sense the whole exercise was pointless, and existence was without purpose. Some would argue that this is how it should be, that life and man himself have come into being in the course of evolution; that man is merely a complex physico-chemical machine, with developed powers of widely distorting the machinery of nature; that in the course of time, the earth will run its course and disappear, and man with it. This is a pragmatic view, not shared by man's recent ancestors with their rather grotesque ideas of magic and religion, nor by a majority of the human race today. Scientists who argue in this way ignore that, however sophisticated, science only observes and measures what is already there. It can show the way to manipulate matter and energy to produce desired effects; it creates nothing and can explain nothing, except that which lies within the realm of a fore-ordained mathematical system.

Science consists of the observation of facts, that is of what happens in an ordered system; the analysis of those facts and the formation of hypotheses arising from them; and the testing of such hypotheses by further observation and mathematical analysis. Technology consists of the application of established facts to the construction of useful devices, most of which in some way or another will affect the environment. If these are the latest trends of ecology, they need not be written off either as damaging or pointless. Indeed, Bishop Berkeley postulated as long ago as 1710 that nothing exists unless there is an 'observer' of it.[1] J. W. Dunne developed the same idea in relation to human consciousness, and the past and future, in *An Experiment with Time* and *The Serial Universe.*[2] If, therefore, one believes that there is any object in existence and consciousness, one must believe also that, sooner or later, an observer of the 'system' must appear.

I say 'of the system' advisedly, because what we can observe is this one mathematical system which rules our lives and destinies and which our senses, aided by the technological aids we can construct, are capable of observing. It is unthinkable, for instance, that miracles are 'paraphysical'; phenomena cannot occur *within this system* which offend against its laws; if they offend against its laws, they are outside the system. In this sense, scientists and mathematicians are rightly sceptical of such phenomena, until they can find an explanation for them. However, 'nature abhors a vacuum', and both scientists and mathematicians appear obsessed with what happens to their mathematical laws when there is a void; at the other end of the scale, they worry about what happens to them at infinity. Religion also worries about the infinite, probably with more reason. This is very odd, because known physical laws show that neither total vacuum nor infinity can exist within our system. The lower end of our scale is represented by absolute zero temperature (0°Kelvin) at which particulate matter loses energy; the upper end is represented by the speed of light at which matter loses its physical properties. Multiply the wavelength of energy waves by their frequency, and you inevitably get the speed of light. What then happens below or above these two points? There may plainly be other mathematical systems with different rules which science is unable to observe or measure. We can only say that they do not exist because we cannot observe them. But can we not observe them? An occult faculty of the human mind could possibly permit the observation of events in a different time scale and account for mystic experiences which it is impossible to ignore but which we cannot explain. Meanwhile, science has only probed a small proportion of the wave range in the mathematical scale, so that much remains to be discovered within our own mathematical milieu.

It may be asked – what has this to do with ecology, earth's environments and the future of the human race? It has, in fact, everything to do with it. Just as there exists a dividing line between the world of politics and science, so there is a dividing line between science and philosophy, including religion. It is important to the future of the world that scientists should realise and admit what many deny, that there could be events which affect us, occurring outside the mathematical system which rules our universe. The possibility of seemingly impossible events could then at least be recognised. If we aspire to control and manipulate our environment, the recognition of such possibilities becomes important. Hesitant steps are being taken to explore so-called extra-sensory perception (ESP), though this is confined to discovering whether it exists or not and not to ascertaining its mechanism. If there is a mathematical plan which requires an observer, then the observer too requires an observer, as was postulated by J.W. Dunne.

Science, to become a more effective force in environment planning, must shed this isolationism. It cannot spread its doctrines if it wraps their meanings in a mystical jargon, unintelligible even to fellow scientists in related disciplines. Incursions of physicists into biology have resulted in unravelling the mysteries of the genetic code. In other ways too, there is a trend away from isolationism amongst scientists, and one hopes to see this trend extend beyond the sciences. Are not finance and economics also subjects susceptible of scientific treatment? Are they not related to the behavioural sciences, and are not the behavioural sciences important to planning of human ecology? Should science shrug off mysticism and religion? In all these ways, one may hope that science in future may better help society to overcome its difficulties. Scientists do not communicate satisfactorily with their fellow men, and their fellow men are dubious of what they have to communicate.

How the present mundane turmoil will end, nobody can foretell. Ecological laws would suggest that we are at the present coming to the apex of a population explosion, following an era of great prosperity. We see around us the signs of the uneasiness and 'stress' that would be expected before the population crash. Cycles of explosion and crash, prosperity followed by decline both of living and moral standards are all too evident throughout human history, though the cycle is not one of years as with lemmings, but hundreds of years. This need not be restated, since it becomes evident even from a rudimentary study of human history and has been operative since Neolithic times. The question arises as to whether nature will take its course, or whether man, aided by his new-found technology, can avert it. Decline would, of itself, cure many of the problems of pollution, inflation, aggressiveness, and feelings of uncertainty, inadequacy and instability. To avoid decline, man must by his own efforts achieve a state of equilibrium with his environment, embracing not only advanced Western communities, but also the more backward communities of the world. The world of man is today a single ecosystem and population must be matched to resources.

In human affairs, population equilibrium is no longer achieved by matching death rate with birth rate; birth rate must be matched with a declining death rate. If birth rate is not matched to the needs of population replacement, then we shall lose our comfortable lives and our affluence, and in some way, which will be unpleasant, nature will do the job for us. This is so well known, that it might be thought superfluous to state it. Yet limitation of births by artificial means is still repugnant to some people on religious grounds, and politicians are surprisingly hesitant to take those steps that are open to them.

Both religious and secular laws admit freedom of choice to the individual over sexual relationships. However, emotive forces are evoked whenever proposals are put forward to make contraceptive advice and devices available to the unmarried. This is illogical, and in any case, once women were emancipated it was, or should have been, obvious that they would claim the same sexual freedom enjoyed by the male, if risks of unwanted pregnancy could be avoided. This development has not led – as it might have done – to a great increase of promiscuity. Most young people establish stable relationships with a partner as an experimental prelude to an association which in most cases has considerable durability. The husband is no longer regarded as the first bedfellow, but the last.

The ecologist may spare himself a quiet chuckle, because such was entirely predictable under ecological laws! 'Pair-bonding', as it is called by ethologists, is deeply ingrained not only in social custom, but in the very nature of the human race. Polygamy has, admittedly, been secondarily developed in some of the races of man, mostly as it would seem from economic causes. In most races from the least to the most advanced, one man and one woman live out their adult lives together sharing the good and the bad, and raising the family. Both make great sacrifices for the sake of the children, in order that they may be well equipped both physically and mentally in the life that lies ahead of them. This does not seem to be incompatible with a certain degree of promiscuity in some cases.

This is quite unlike what happens with our nearest relatives the anthropoid apes, except for the gibbons. Chimpanzees, in particular, are quite promiscuous; the female mates with a number of males when at oestrus, and there is no pair-bonding. The tendency to monogamy in man must have had survival value during the period when his ape-like ancestors were forced to leave the trees and adopt a terrestrial mode of life. The advantage must have lain in the preservation of the family unit, in the teaching of love and duty, and of useful skills to the young. The value of the family as a unit was never more clearly shown than today, when the unit is no longer so binding as formerly, and it is the children from broken families that provide most of our bad citizens and delinquents. Evidently also, sex and the performance of sex between partners came to be an important factor in cementing the pair bond between male and female. In no other primate is there the same marked difference between the sexes, a difference which leads to such an intense mutual desire for each other of one *particular* man and one *particular* woman, which adds up to what we call 'love' and leads to the stable relationship. How many times do the selected pair mate for each child born? Plainly, sex is as important socially in human affairs as for reproduction. This must have come about at least 100,000, if not a million, years ago.

It must be asked, when crystal-gazing into the ecological future, whether the pair-bonding system retains its value and will survive, or will it lose its value for urbanised societies and disappear? It is something which man and woman still plainly value, and most wish to retain. Nevertheless, promiscuity and broken marriages are on the increase, and these could indicate that an increasing proportion of urbanised society no longer values the pair-bond link, and would prefer a free-for-all. With family planning, fewer children are born, and over a shorter period, so that the family unit might come to have less value for their welfare.

It is fashionable amongst prophets of doom to paint horrifying pictures of the present-day breakdown of society and standards of behaviour. However, analysis of these trends hardly bears out these gloomy prognostications. A study of human history since the Christian era shows how much more tolerant and better off we are than ever before. Education and medicine are available to all. Every boy and every girl born into Western society has at least a reasonable chance of living his or her natural life span in health and contributing to human welfare. We do not kill the first born nor sacrifice our enemies or unfortunate virgins to the gods. Perhaps we have advanced too far in tolerance. If so, this is a revulsion against the wars and barbarities of the past. Some groups and individuals do take advantage of this tolerance and often punishment does not seem to fit the crime. What will be the effect on society as a whole of a failure to eliminate anti-social groups or persons? Society is at present seeking the answer. Crime and addictive habits are regarded as diseases requiring therapy rather than revenge. Obviously, the human animal is still badly adjusted to urban environments; ecology and evolution may have the last word in promoting greater adjustment over the generations. Adjustment of population levels, of behaviour patterns, and of general stability would be expected, if ecological laws continued to operate and man's environment continued to be sufficiently stable. As we saw in the last chapter, evolution still seems to operate on *Homo sapiens.*

The most frightening aspect of ecology is the total failure of world leaders, both national and international in political and economic spheres, and their inability to understand the simplest facts of science and technology. Political science must match physical science, if a stable environment for earth's inhabitants is to be planned. But, will a new type of planned environment remain stable even if it is achieved? Stability depends on two sets of factors: first, those that man can control; secondly, those beyond human control. Undoubtedly, man can adapt his environment to achieve equilibrium and stability. Whether he is temperamentally capable of doing so is another question and may be doubted. If he does not do so, he will either disappear into biological history, or enter a new dark age to re-emerge in a thousand years or so, and repeat the attempt. If successful, man faces a number of challenges in the technological sphere in the not-too-distant future. His comfortable life is attained by living on capital, by the use of fossil fuels laid down on earth in former biological ages. The greatest challenge to his technology is imminent, when he must find the means to tap alternative sources of power, sources that are income and not capital, and will, therefore, be lasting. The use of atomic energy presents many problems, and possibly the only lasting source of power will prove to be that of the sun trapped outside the earth's atmosphere, that power which originally gave birth to life itself. This possibility is, of itself, a justification for the exploration of space techniques.

Of the influences on human history of the future that lie outside the powers of man to manipulate, the most serious is that of climatic change. We do not know whether our present climate represents the end of the last ice age, or an intermission. If the latter, the glaciers will again descend from the north with a force and inevitability that no human effort can counter. Judging from glacial cycles, this could occur in 5,000 to 10,000 years from now. The northern hemisphere could be glaciated almost as far as the Mediterranean and the Gulf of Mexico. Floods would occur in the tropics and the Sahara would become well watered and fertile. 'The best laid plans of mice and men gang aft agley.'

As an epilogue to this work, it will be advisable to consider the earth as a single ecosystem of the human biosphere. If the earth is to be regarded as such, problems of population, food production, pollution, finance and economics are incapable of solution, unless mankind as a whole acts in unison, and unless world leaders of the necessary stature appear. At present, the outlook for such is gloomy. The peoples of the world are divided into warring factions. The Western civilisations are tensed, uneasy and stressed amidst unprecedented prosperity; 'iron-curtain' countries are bellicose; the rest of the world is undernourished and impoverished. How can the uniformity required by a single ecosystem come into existence under such conditions? Geneticists have been vocal on the subject of race differences and their responsibility for the existence of these conditions. Strangely, ecologists have been silent on the subject. So let us examine the question against an ecological background and in relation to two main topics: first, the problem of immigrant alien races; secondly, the differing abilities of the races.

Both zoology and ecology show that conflict arises when there is immigration of an allied but different race or species of animal into an established habitat. We have already studied the teachings of Alfred Russell Wallace of the effects on each other of races developed in isolation. It should not surprise us, therefore, when the same occurs in human habitats. Whites are resented in black areas, and blacks in white areas. Jews are resented by their cousins, Semites, the Arabs. The trouble is less pronounced when the two races occupy different 'niches' in the habitat. To a certain extent, this happens with the Jews in Western societies, with their skills in trade and finance. However, their ability to accumulate wealth arouses resentment which at times boils over. Coloured peoples tend to occupy the niche of menial work, and they naturally resent their exclusion from more skilled jobs. However, if they emerge from this niche, they invade the niche of the white worker and are in their turn resented. This is in conformity with the ecological 'Principle of Competitive Exclusion', which states that two dissimilar races cannot occupy the same niche in the habitat. Those who ask for multi-racial states, in which the different races share equally, ask for the suspension of this natural law. Man has defied ecological laws before, and perhaps he can do so again. However, I cannot see Black Africa accepting while immigrants, or Indians or Chinese for that matter, on an equal basis with its own nationals, even if some semblance of integration were achieved in predominantly white countries.

If different niches cannot be found for competing races, ecological laws demand one of three other solutions:

1. The immigrant population will replace the existing one, of which many instances can be cited in human history;
2. The existing population will drive out the immigrant, as eventually the Spaniards did to the Moors;
3. The immigrant population will be assimilated.

The most recent example of assimilation is that of the United States in which different white races have been welded into a new nation. In countries such as Brazil, the coloured peoples have also been assimilated with a high degree of success. Britain, too, has assimilated waves of refugees and trading communities, who after a generation or two are more British than the British. With races such as the Jews, dedicated to preserving a national and religious identity, and coloured peoples, who have marked physical differences, assimilation becomes much more difficult. In the course of time, probably a long time, possibly they too might be assimilated, and this may be happening in the United States where white blood is much diluted by black. It has also been happening for many generations in parts of Africa, such as coastal Ghana, where black blood is much diluted by white. However, the problem is made more acute because of the widespread belief amongst whites that the black races are mentally inferior. This is the second problem which should be examined from the ecological viewpoint.

Scientists are often not particularly convincing when they stray from their chosen field. This remark applies especially to geneticists, when they stray into the realms of sociology. On the basis of purely sociological intelligence tests, some geneticists claim that black races are intellectually inferior to white in respect of IQ. Others maintain that the apparent difference is not genetical, but the result of social disadvantage and a poor background in childhood. It would appear from the evidence that both are partially right, but in any case from an ecological point of view the argument is completely sterile and argued on the wrong bases. It is further said that persons of any race with an IQ of under 100 should be given financial inducements to be sterilised. So, we are all to become Einsteins, and there will be no one to deliver the coal! This sort of argument is as dangerous as the proposal to alter genetic constitution by so-called 'genetic engineering'. Moreover, any such procedure would certainly not have the desired effect.

If we accept that man is a single species within the genus *Homo*, *H. sapiens,* and this we do, then all races share the same range of genes. It follows that differences of race, both mental and physical, are due not to gene *differences* but to a different distribution and combination of the same genes within the human 'polymorphic genetical system'. This is a very different thing. Man's 'gene pool' is enormously wide, and it is to this wide gene pool that he owes the many varied abilities which go to make his culture. His culture is dependent on a complex social structure, which in turn is dependent on the wide stretch of abilities which the wide gene pool makes possible. During the long and difficult centuries of the ice ages and after, evidently there was a genetical discrimination in the white races in favour of selecting and coupling those genes which make for higher intelligence. Undeniably, the white and other northern races have proved the more inventive and more capable of manipulating an unfriendly environment.

Faced with the challenge posed by Western civilisation and urbanisation, one would expect that a recoupling of genes would rapidly take effect in those races which had not experienced this particular type of challenge previously. No fundamental genetic change, as by mutation, is involved, merely a reshuffling of existing genes. Anybody who breeds animals knows how quickly new races or varieties can be established. IQ tests, therefore, whatever they show or prove, are totally irrelevant to the problem, though it is silly to suggest, as some do, that the problem does not exist. The ecological response in backward races to a changed environment should be adaptation to it, within a comparatively small number of generations.

Ecological adaptation is not altogether painless. While the polymorphic genetical system will work for a greater spread of abilities and increased powers of leadership and performance at one end of the scale, it also leaves a trail of inadequates at the other. This phase of the operation is evidently particularly active at the present time amongst urbanised Western communities. As we have already seen, these unfortunates tend to be 'genetically unfit' and so in the long run discarded. There might also come a time in human history when physical endurance was more important to survival than high intelligence. In this event, IQ would recede and the robuster individuals would be the survivors. It is this flexibility of the polymorphic genetical system which is of such importance. A change of gene structure by mutation would take thousands, even tens of thousands, of years to alter race characteristics, against a few generations by gene reshuffling.

Every gene has one factor from the father and one from the mother. These may be dissimilar, and each gene is recombined at each generation. Moreover, many characters, particularly perhaps intelligence, are determined by more than one gene. Furthermore, there are other 'modifier' genes, which modify genes' effects. So to talk of selecting human beings for intelligence is utter nonsense; only the challenge of the environment can do this. Wipe out a country's intelligentsia, as has happened in France and Russia; within a generation or two, equally gifted leaders have arisen. At any rate, how profoundly foolish it would be to select man for one or two characters, when his social structure demands such a wide spread of abilities and his survival is dependent on the ability of his genetical system to adapt the physical characters of the race quickly to changed environmental circumstances. In general, white races have become adapted to life in temperate climates. They are not readily adapted to life in tropical regions; while the first generation performs effectively, the second and succeeding generations decline in vigour. Conversely, the black races have become adapted to life in hot countries, and in the absence of refreshment by newer immigration it is doubtful whether their numbers would be maintained in temperate climates. During the sixteenth and seventeenth centuries, large numbers of Africans had been brought to Britain; by 1939, they had disappeared. Possibly they were absorbed into the general population; possibly they declined because of greater susceptibility to infectious diseases, such as tuberculosis and pneumonia, or to nutritional diseases such as rickets. During my travels in the United States, I have found the negro people more contented and more in tune with the environment in the south than in the north, in spite of supposed greater intolerance there.

The lessons of ecology would appear to suggest that the different races will perform best in the areas which have produced them. In such areas, I do not doubt that the backward peoples, in response to the change of the environment, will quickly become equipped with the same range of talents that has appeared so quickly in Western and Eastern cultures. The signs are already to be seen in African universities.

It is, in any case, profoundly to be hoped that this forecast is correct. The world, as a single ecosystem of the human biosphere, cannot afford two standards of culture, the rich nations and the poor nations, any more than Victorian England could afford Disraeli's two nations. Ecological disequilibrium means instability. It seems to me that the solution of environmental problems is dependent, more than anything else, on the ability of the races of man to develop their maximum potential within their own niches and in harmony with each other, so that world problems can be tackled jointly.

NOTES

1 Bishop G. Berkeley, *Principles of Human Knowledge,* 1710
2 J.W. Dunne, *An Experiment with Time,* Faber, London, 5th edn., 1939, and *The Serial Universe,* Faber, London, 1934

Further Reading

Carthy, J.D. and Ebling, F.J. (eds), *The Natural History of Aggression,* Inst. of Biol. Symposium, Academic Press, London & New York, 1964
Montagu, A., *Human Heredity,* Signet Science Library, New York, 1963
Parlin, C., *The Relations between the Sciences,* University Press, Cambridge, 1968
Wolfgang, W., *The Sexual Code. The Social Behaviour of Animals and Men,* Doubleday, New York, 1972

INDEX